# FACTS FROM SPACE!

---

## FROM SUPER-SECRET SPACECRAFT TO VOLCANOES IN OUTER SPACE, EXTRATERRESTRIAL FACTS TO BLOW YOUR MIND!

---

*Dean Regas*

ADAMS MEDIA

NEW YORK   LONDON   TORONTO   SYDNEY   NEW DELHI

**A**adamsmedia

Adams Media
An Imprint of Simon & Schuster, Inc.
57 Littlefield Street
Avon, Massachusetts 02322

For information about special discounts for bulk purchases, please contact Simon & Schuster Special Sales at 1-866-506-1949 or business@simonandschuster.com.

The Simon & Schuster Speakers Bureau can bring authors to your live event. For more information or to book an event contact the Simon & Schuster Speakers Bureau at 1-866-248-3049 or visit our website at www.simonspeakers.com.

Interior images © Clipart.com; iStockphoto.com.

Manufactured in the United States of America

10 9 8 7 6 5 4

Library of Congress Cataloging-in-Publication Data
Regas, Dean, author.
Facts from space! / Dean Regas.
Avon, Massachusetts: Adams Media, [2016]
LCCN 2016020321 (print) | LCCN 2016024632 (ebook) | ISBN 9781440597015 (pb) | ISBN 1440597014 (pb) | ISBN 9781440597022 (ebook) | ISBN 1440597022 (ebook)
LCSH: Astronomy--Popular works. | Solar system--Popular works. | Cosmology--Popular works.
LCC QB44.3 .R44 2016 (print) | LCC QB44.3 (ebook) | DDC 523.2--dc23
LC record available at https://lccn.loc.gov/2016020321

ISBN 978-1-4405-9701-5
ISBN 978-1-4405-9702-2 (ebook)

# CONTENTS

# INTRODUCTION

Did you ever want to experience the microgravity of space? See what's hiding behind the clouds that shroud the surface of Venus? Learn what happens inside a nebula when a star is born?

If you've ever wanted to learn more facts about space, you're in the right place—and you're not alone! When I was ten years old and saw pictures of the beautiful ringed planet Saturn, I vowed that someday I would fly to Saturn and ride my bike around those rings. I didn't want to be an astronomer back then. I wanted to be an adventurer. My ten-year-old self dreamed of visiting far-off lands, and outer space was the farthest thing I could think of.

I became an astronomer instead of an adventurer, but with the facts in this book, I'm still going to take you to newly discovered worlds. Here you'll find facts about astronomical objects that will make you pause and see the universe in a whole new way. Filled with the accumulated knowl-edge of thousands of years of watching the skies, you'll learn all you need to know about what it would be like to watch a Martian sunset, to swirl in Jupiter's Great Red Spot, and skim the corona of the Sun. Your imagination will soar among the stars, see constellations from different perspectives, fall into a black hole, and more.

Every day my colleagues discover an asteroid in the solar system, a planet in another star system, a star or star cluster in the Milky Way, or even an entire, never-before-seen galaxy in the universe. Even PhD astrophysicists who have studied the universe their entire lives get their minds blown on a regular basis. And that is what makes astronomy such a great adventure. So grab this book and get ready to take a ride around the rings of Saturn and beyond—3-2-1 blast off!

# CHAPTER 1

# ORBITING EARTH

*Tales of Humans (and Nonhumans) in Space*

Space travel is not glamorous. Imagine being confined to your cabin on a cruise ship on a long ocean voyage, or traveling for a week in a bus that you can never leave. This will give you an idea of the conditions astronauts endure living in the final frontier. It's claustrophobic, it's dirty, and the food is tasteless.

Anyway, you're weightless, which may seem fun at first, but it can actually cause nausea, sleep problems, sinus congestion, muscle loss, a puffy face, and wild and uncontrollable hair. And that doesn't even take into consideration the unmentionable bodily fluids that float around you—that you probably don't want to float around you.

You are rewarded for all this inconvenience and discomfort with one heck of a view. In orbit you would see the entire planet from a satellite's viewpoint. You'd see a lot of blue: oceans, seas, lakes, and rivers. You could marvel at mountains, islands, as well as manmade wonders. You can watch cities from above and observe how they light up at night. You'd experience enough sunrises and sunsets to last a lifetime. You'd go where few have gone before—aside from a select few astronauts of various ethnicities, a handful of dogs, and some literal guinea pigs. What an adventure!

First let's look at the telescopic advances that led us to dream of reaching the stars. Then we'll blast off to orbit the Earth and delve into what living in near weightlessness is like—the good, the bad, and the ugly.

# TELESCOPES AND OBSERVATORIES

## THROUGH THE LOOKING GLASS

On September 25, 1608, a Dutch spectacle maker named Hans Lipperhey applied for a peculiar patent. About his invention Lipperhey wrote, "All things at a very great distance can be seen as if they were nearby, by looking through glasses . . ." Although lenses and eyeglasses had been manufactured since the fourteenth century, Lipperhey combined two lenses to create the first documented telescope.

## SIGHT UNSEEN

By 1609, telescopes were on sale in Paris, demonstrated in Germany and Italy, and had attracted the attention of Galileo. After hearing the scantest details about such a device, Galileo made his own telescope in less than twenty-four hours. Galileo greatly improved on the design and began pointing the scope skyward. He documented the craters of the Moon, the phases of Venus, the moons of Jupiter, and even spots on the Sun. He then observed never-before-seen stars and published his observations.

## MONSTER TELESCOPES

The larger the telescope, the better the view. And so a new generation of astronomers began making their telescopes bigger—some reaching gigantic proportions. In the mid-1600s, Dutch astronomer Christiaan Huygens worked with his brother Constantijn to construct a tubeless telescope in which the two lenses were separated by 123 feet. Polish astronomer Johannes Hevelius topped that with a 150-foot-long monstrosity that he claimed was easy to use (with the help of a crew of assistants operating the ropes and pulleys).

## AS BIG AS THEY COME

The use of the lenses that gave refracting telescopes their name also meant that they had a limited size. If the lenses were too large, their weight could not be adequately supported. The largest refractor in the world is forty inches in diameter and is still used at the Yerkes Observatory on Geneva Lake in Williams Bay, Wisconsin.

## LET'S REFLECT

English scientist Isaac Newton revolutionized telescope design by producing the first usable reflecting telescope in 1668. Instead of using lenses, Newton positioned two mirrors to reflect the image into the glass eyepiece. The design made telescopes more compact and had no size limit.

## MORE POWER!

In the eighteenth century, English astronomer William Herschel constructed a twenty-foot-long reflector and then a forty footer, which looked like huge cannons mounted on pivoting scaffolding. The diameter of the objective (the lens or mirror) is the biggest factor in a telescope's power. After Herschel's designs, telescopes were then measured by their diameter, not their length. Astronomer William Parsons (also known as the 3rd Earl of Rosse) constructed a massive scope in Ireland dubbed "The Leviathan of Parsonstown," which had a whopping seventy-two-inch-diameter mirror in 1845.

## DROPPING THE BALL

In the nineteenth and twentieth centuries, observatories used to be the official timekeepers for cities. To better inform the public, observatories set up time balls that they would drop at noon each day. People could look up and see the ball drop and know what time it was. Starting in 1907, New Yorkers made a ball, very similar to those used by observatories, with one addition: lights. On New Year's Eve 1907, they dropped the ball in Times Square, and it has been a tradition ever since.

## BIGGER AND BETTER

Some modern optical telescopes use a series of segmented mirrors to gather light from the cosmos. The twin Keck telescopes perched atop Mauna Kea, Hawaii, employ thirty-six hexagonal mirrors that measure, in total, 394 inches across. Although larger telescopes are being developed, the Gran Telescopio Canarias located in the Canary Islands can gather the most light with an aperture of 409 inches.

## ON ANOTHER WAVELENGTH

Some telescopes capture wavelengths of energy beyond visible light such as microwave, ultraviolet, and infrared light. The largest telescopes pick up sources of radio waves with large parabolic dishes. The Arecibo telescope utilizes a 1,001-foot dish that is built into a depression in the mountains of Puerto Rico. Although its dish is smaller than Arecibo's, the Green Bank Telescope (GBT) in West Virginia is the largest movable object on land. Weighing about 8,500 tons, the GBT can point to almost any object above the adjacent Appalachian Mountains.

## SMALL BUT SPECIAL

The Hubble Space Telescope (HST) is a small telescope by comparison—its mirror is only ninety-four inches in diameter—but it is still performing some magical feats of imagery. Launched in 1990, HST does not have to worry about clouds, fog, rain, or pollution since it orbits 150 miles above Earth. From its lofty perch above, HST's mirror and cameras have peered to the farthest reaches of the universe.

# UP, UP, AND AWAY!

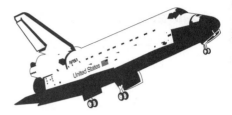

## NASA KNOWS SPACE

*NASA* stands for the National Aeronautics and Space Administration. Formed in 1958, this government agency coordinates and runs manned and unmanned space missions and aerospace research in America.

## WHERE DOES "SPACE" BEGIN?

While it's tough to say exactly where Earth's atmosphere ends and outer space begins, American astronauts consider anything beyond sixty-two miles from Earth as "space." For reference, commercial jets fly no more than ten miles up. Sorry frequent fliers, you haven't quite been into space.

## SHORT RIDE UP

If you blasted off from Earth in a rocket designed for manned missions, you'd reach space in about two minutes and thirty seconds. If you cut the engines off at that point, you still wouldn't be going fast enough to circle Earth, and you'd fall back to Earth in pretty short order.

## ZERO TO SIXTY IN 1.75 SECONDS

To orbit Earth, a spacecraft must reach a speed of at least 17,400 miles per hour. It would take you about eight and a half minutes to achieve the proper altitude and velocity. Think about that: going from 0 to 17,400 miles per hour in 510 seconds. That means you would accelerate 60 miles per hour every 1.75 seconds.

## SIT BACK AND RELAX

During launch, astronauts lie back in their chairs in order to absorb the tremendous forces caused by this acceleration. This force, called g-force, is expressed with respect to the gravity of Earth, with 1 g being the normal gravitational force you feel on Earth. Astronauts typically experience 3 gs during takeoff, which makes their bodies feel three times heavier than normal. Their reclined takeoff position is called "eyeballs in" because it feels like their eyeballs are going into their heads.

## NEARLY WEIGHTLESS

Once a spacecraft achieves its orbit around Earth, the engines shut off and you become weightless—well, almost. You are actually in microgravity, or where g-forces are almost zero. You are still being pulled toward Earth and you are in constant freefall. However, you are traveling so fast around Earth that you never hit the ground. Essentially, you keep falling and missing Earth.

## COSMIC ROAD TRIP

The International Space Station (ISS) orbits between 205 and 255 miles above Earth, which is roughly the driving distance from Cleveland to Cincinnati, Ohio. The trip is so short that astronauts probably don't have to ask, "Are we there yet?"

# ANIMALS IN SPACE

### FEARLESS, FOUR-LEGGED FLYERS

Who was first in space? John Glenn? Alan Shepard? Yuri Gagarin? Nope. Laika, a dog. Before any human reached the final frontier, a troop of animal astronauts in the 1950s and 1960s pioneered space flight. Dogs, rats, mice, monkeys, turtles, and guinea pigs led the way. The first astronauts walked on four legs—and most of them were Russian.

### UPS . . . AND DOWNS

In the 1950s, the Soviet Union had a clear edge in the space race. Between 1951 and 1960, the Soviets launched more than a dozen high-altitude test flights with animals as passengers. These rockets flew suborbital missions—meaning they made it into space but fell right down again without circling the globe.

### FREQUENT FALLER

The most well-traveled suborbital passenger (who never circled the globe) was the dog Otvazhnaya—"Brave One" in Russian—who made five successful flights between 1959 and 1960. She shared one of the flights with another dog named Snezhinka (Snowflake) and a rabbit named Marfusha (Little Martha).

### SPACE STRAYS

Russian scientists preferred working with dogs in their space missions. They rounded up small, female stray dogs from Moscow streets and brought them to the research center for training. One minute these dogs were wandering the alleys of Moscow, and the next they were on their way to the stars.

## DOGONAUTS

The Soviets believed that dogs were better suited for space flight than other animals. They had the ability to endure extensive periods of inactivity in an enclosed space—a must for all space travelers even today. And while in flight, a dog could wear a pressurized suit and helmet.

## NO FIRE HYDRANT NEEDED

The Soviets preferred to send female dogs to space due to their ability to urinate in any position. Engineers actually designed spacesuits specifically for female canines, and there was no room in the cabins for the dogs to raise one leg.

## LAIKA'S TRAGIC FATE

On November 3, 1957, the Soviets launched Sputnik 2, which carried the first passenger into space. Space dog Laika ("Barker" in Russian) captured the attention of people around the world. Some American newspapers reported that "Prayers were said for the dog and people were asked to observe a minute's silence each day 'with special thoughts for her early and safe return to Earth.'" Unfortunately, Soviet scientists failed to make any provisions for Laika's safe return. She orbited Earth only a few times before the cabin became overheated and she died. On April 14, 1958, the spacecraft fell into Earth's atmosphere and burned up.

## ANIMAL RIGHTS ON EARTH

The safety of the early Soviet missions was highly questionable. Out of thirteen dogs sent into space, five never returned. In particular, Laika's mission to space sparked a renewed debate in Russia, America, and around the world regarding the ethical treatment of animals in scientific research.

## INTERSTELLAR REGRET

In 1998, Oleg Gazenko, the former Soviet scientist from the animals-in-space program, expressed regret regarding the fact that space dog Laika died in space. During a Moscow news conference he said, "The more time passes, the more I'm sorry about it. We shouldn't have done it. . . . We did not learn enough from this mission to justify the death of the dog."

## THEY STUCK THE LANDING

The first truly successful space mission was Sputnik 5. Two dogs, named Belka and Strelka, made the arduous journey along with forty mice, two rats, and a variety of plants. The crowded spacecraft completed eighteen orbits of Earth and returned all creatures safely to Earth. Strelka later gave birth to six healthy puppies—one of which was given to President John F. Kennedy as a token of friendship.

## HIGH-FLYING HOUNDS

The record for canine courage and stray-dog stamina belongs to Veterok (Little Wind) and Ugolyok (Coal). This pair remained in space for twenty-two days—a record that lasted until NASA's 1973 Skylab 2 mission.

## THE ZOO ABOVE THE MOON

In 1968, a Soviet spacecraft named Zond 5 carried turtles, flies, mealworms, and bacteria on a seven-day trip around the Moon, three months before humans tried it. After Zond's splashdown in the Indian Ocean, all living creatures were safely recovered. Although the turtles lost 10 percent of their body weight, they soon returned to normal turtle existence on Earth.

## MONKEYNAUTS

While the Soviets preferred to use dogs in their space missions, the Americans sent monkeys to test outer space. The majority of these flights were suborbital, meaning the rockets just shot up to around 300 miles and fell back to the ocean without circling the globe. They were not much more successful than the Soviet attempts. Four out of ten primates died during these missions in the 1960s.

## FLOTATION FATALITY

The first monkeynaut was a squirrel monkey named Gordo who flew atop a Jupiter rocket in 1958. Gordo survived the flight and the splash-down into the Atlantic Ocean. Unfortunately, the craft's flotation devices did not engage, and neither Gordo nor the spaceship were recovered.

## PRIMAL PRIMATES

Primates named Able, Miss Baker, Sam, Miss Sam, and Ham made successful suborbital flights in 1959 and 1960. But it wasn't until November 1961 that an American animal orbited Earth. Enos the chimp made two orbits of Earth as the prelude to John Glenn's flight three months later.

# HUMANS IN SPACE

## SOVIET SUPERSTAR

On April 12, 1961, Soviet cosmonaut Yuri Gagarin became the first person in space and the first person to orbit Earth. Every year on April 12 is Yuri's Night, when people around the world celebrate this achievement and throw space parties in his honor.

## PHONE HOME

Just before liftoff, Soviet cosmonaut Yuri Gagarin exclaimed, "Поехали!," which translates as "Let's go!" When he landed in a field more than 100 miles off target, the first people he saw were a farmer and daughter. Wearing a bright orange spacesuit and dragging his parachute behind him, Gagarin had to reassure them that he was their comrade and said, "I am a Soviet citizen like you, who has descended from space and I must find a telephone to call Moscow!"

## SENIORS IN SPACE

When John Glenn was forty years old, he was the first American to orbit Earth, in the *Friendship 7* capsule. When he flew again aboard the space shuttle *Discovery* in 1998 at the age of seventy-seven, he became the oldest person to orbit Earth. On his first trip to space, upon first feeling the effects of weightlessness, he said, "Zero G, and I feel fine."

## NOT MADE FOR WALKING

When astronauts in space put on a spacesuit and leave their craft, it's called an extravehicular activity (EVA). It's more commonly referred to as a "space walk," although without any ground under your feet you wouldn't do much walking.

## SPACE WALK SNAFU

The first EVA was performed by Soviet cosmonaut Alexey Leonov in 1965. Connected by a tether, Leonov left his Voskhod 2 spacecraft and floated outside for twelve minutes. He had trouble getting back inside Voskhod 2 however when his suit greatly inflated. He didn't fit! Leonov had to depressurize his suit so that he could move his arms and finally reach safety. During the EVA, Leonov perspired so profusely inside the suit that he described being up to his knees in sweat.

## UNHAPPY LANDING

Due to their bulky spacesuits, the crew of Voskhod 2, Leonov and his commander Pavel Belyayev, could not settle into their seats before landing. This threw off the center of mass of the spacecraft, which then entered Earth's atmosphere off course. The cosmonauts landed in an uninhabited forest and had to spend one night in their capsule and another in a hut they built with skiers who came to rescue them before catching a helicopter ride back home.

## BE PREPARED

The cosmonauts onboard Voskhod 2 had a pistol and ammunition. The crew reported finding this very reassuring since they were likely to come into contact with bears or wolves—denizens of the taiga forest in which they landed.

## STOWAWAY SANDWICH

American astronaut John Young tucked a corned-beef sandwich into his spacesuit and smuggled it aboard his 1965 Gemini 3 mission. He and his crewmate, Gus Grissom, enjoyed a few bites instead of their normal space food. Young was mildly disciplined by NASA officials for the contraband sandwich when he got back home.

## JINGLE BELL JOKES

There was eerie music coming from the Gemini 6A spacecraft. Command pilot Wally Schirra produced a harmonica that he stowed onboard. After claiming that the sounds came from a UFO, Schirra proceeded to play "Jingle Bells" on the harmonica. This seemed appropriate since the Gemini 6A mission happened in December 1965.

## LIGHTNING STRIKES TWICE

In 1969, as the Saturn V rocket carrying the Apollo 12 crew blasted into space, it was struck by lightning—twice. Despite the damage this caused to fuel cells and guidance systems, NASA determined that the spacecraft was good to go. And indeed it was! The spacecraft went on to the Moon and then returned the three crewmates safely to Earth.

## SPACE IS LIKE A BOX OF CHOCOLATES

In 1978, while traveling to the Salyut 6 space station, Soviet cosmonaut Aleksandr Ivanchenkov found a box of chocolates that his wife smuggled onboard. When he opened the box, chocolates flew everywhere. It took him two hours to collect (and/or eat) them all.

## NEARLY DROWNED IN SPACE

Italian astronaut Luca Parmitano nearly drowned during his space walk in 2013. He felt cold water inside of his suit, and soon it started filling his helmet. Parmitano did not panic but made it safely back inside the ship. NASA is still not sure what went wrong to cause his suit to malfunction like it did.

## WRONG NUMBER

The International Space Station (ISS) has a phone. In 2015, British astronaut Tim Peake used it but unfortunately dialed the wrong number. When he asked, "Hello, is this planet Earth?" the woman on the other end of the line assumed it was a prank call. On another occasion, Peake tried calling his parents from the ISS but just missed them. He had to leave a message on their answering machine.

## SPACE STATION VACANCY: ASTRONAUTS WANTED

There have been many orbiting space stations prior to the ISS. The first one was the Salyut I launched by the Soviet Union that circled Earth for 175 days. The American space station, Skylab, orbited for over 2,000 days but was only occupied for 171 of them.

## MIR CIRCLES THE WORLD

The Mir space station launched in 1986, before the fall of the Soviet Union, but Russia maintained it through 2001 when it fell back to Earth. Cosmonauts occupied Mir for more than 4,500 days.

## DOES THIS THING HAVE A GUEST ROOM?

The ISS was a joint effort between the United States, Russia, and several other countries. It went up into space, piece by piece, starting in 1998. It now weighs more than 900,000 pounds and is 239 feet wide, 356 feet long, and 66 feet tall. Six crew members can live there at a time, but there is enough room to entertain guests (and the replacement crew).

## SUNRISE, SUNSET

Traveling at more than 17,000 miles per hour, the ISS circles the entire planet in a little over ninety minutes. That means every ninety minutes the astronauts experience a sunrise—and in between witness just as many sunsets.

## EXCLUSIVE ACCOMMODATIONS

In 2011, China made its own space station named Tiangong-1. Until now, Tiangong-1 has only been occupied for short periods of time. This is the first phase and first module of a larger, planned space station that will hold a crew of three starting around 2020.

## SHORT-SLEEVE ENVIRONMENT

In spite of the coldness of space and heat of solar radiation, the temperature in the cabins of manned spacecraft is kept around 70°F. When a craft is bathed in sunshine, the cabin must be cooled. But at night, you'll need the heaters.

## THE UNTWINKLING STARS

Astronauts can see the stars by day or night, but from above Earth's atmosphere, sunlight does not scatter and stars do not twinkle. They appear instead as pinpoints of unwavering light.

## A FAMILIAR SKY

Astronauts are not significantly closer to the stars. Since they are only a few hundred miles above Earth, and stars are trillions of miles away, the stars look the same size. Additionally, astronauts see the exact same star patterns and constellations that you see from Earth.

## NO ONE CAN HEAR YOU SCREAM

There is no air in space and therefore no sound. You can only hear astronauts talking through their communication devices because they are in air-filled spacesuits. Sci-fi action movies would be a lot less thrilling if they accurately portrayed soundless laser battles and explosions.

## YOU MUST BE THIS TALL TO RIDE THIS ROCKET

At six feet four inches, Jim Wetherbee was the tallest person to fly in space. Nancy Currie was the shortest at five feet tall. American astronauts must be between fifty-eight-and-a-half and seventy-six inches tall to fly in a NASA mission.

## THE SHORTER THEY ARE . . .

Soviet cosmonaut, and first person in space, Yuri Gagarin stood five feet two inches tall. For long space missions, shorter astronauts might have a distinct advantage. They take up less space, consume less food, and expel less waste.

## WHEN YOU GOTTA GO

It's a fact that space toilets are not fun to use. With no "down" in space, no strong pull of gravity, how do you keep things you don't want to see again from floating with you? One word: vacuums.

## TMI

The space toilet aboard the space shuttles broke during two missions. One was repaired in flight, while the other could not be fixed. The poor astronauts had to use the backup system, called the fecal containment device: basically a bag they taped to themselves.

## CAN'T YOU HOLD IT?

Before becoming the first American to reach space, Alan Shepard sat suited up in the rocket on the launch pad, ready to fire, for way too long. He was faced with an urgent problem. He told Mission Control, "I've got to pee." Instead of stopping the countdown, Mission Control gave Shepard permission to go in the suit. Luckily his flight aboard the Freedom 7 Mercury capsule into space was only a brief, damp fifteen minutes and twenty-two seconds.

## SPACE DIAPERS

Because of possible delays, the rigors of liftoff, and the fact that space-suits don't have zippers, astronauts wear Maximum Absorbency Garments (MAGs), which is a fancy way to say "space diapers." Astronauts also wear MAGs whenever they leave the ship to perform extravehicular activities (EVAs). These incredibly engineered diapers can hold up to two liters of fluid, which is a good thing because there's no way to take a potty break when you're on a space walk.

## PICTURESQUE PEE AND MORE

A spare glove dropped by American astronaut Ed White during the 1965 Gemini 4 mission circled Earth for months before re-entering Earth's atmosphere and burning up. However, the most startling thing left in space is pee. Early space missions ejected human waste into space where it froze into beautiful patterns.

## REDUCE, REUSE, RECYCLE

If you expose human waste to the vacuum of space, all the bacteria die—along with the smell. On the International Space Station, urine is purified and recycled into fresh, clean drinking water.

## SPACE LASERS WORK

Humans have left roughly 500,000 items in space. Discarded waste, parts of rockets, and broken or obsolete satellites litter the lower reaches of space where NASA monitors them. And it's a good thing, too! If a rocket flying 18,000 miles per hour runs into a stray screw left behind from some bygone mission, it could be disastrous. Scientists are researching ways to arm future spacecraft with high-powered lasers that would either vaporize space junk or shoot it down to Earth where it will burn up in the atmosphere.

## NIGHTLY VISITORS

You can see several satellites with the naked eye circling overhead every night. They appear as slow-moving, nonblinking lights that take about five to six minutes to go from horizon to horizon. Tiangong-1 looks like a semibright star, but the ISS can be almost as bright as the planet Venus. When you see it, be sure to wave at the astronauts inside that are about 200 miles above you.

## IRIDIUM FLARES

The brightest manmade objects that you can see with your naked eye are iridium satellites. These sixty-six communications satellites have highly reflective solar panels, so when the sunlight hits them squarely, a beam of light flashes toward your eye for a brief moment. Amateur astronomers call these iridium flares, and these flares can be dozens of times brighter than Venus. If you're not expecting to see one, it is so startling that you may think it is a UFO.

## AMATEUR ASTRONOMERS HAVE AN EYE ON THE SKY

You can't hide things in space from amateur astronomers. A secret, unmanned, reusable U.S. Air Force spacecraft, nicknamed X-37B, that circles the globe was first observed by amateur astronomers in 2010 when no one else knew about it. What is X-37B doing? The Air Force gives very few details. Amateur astronomers continue to keep tabs on it and publish its path online, so even if you don't know what it's doing, you can know where and when it flies.

## FORGET WATER COOLER CHAT

Above Earth's protective atmosphere, astronauts are exposed to more solar radiation than on the ground. Shielding astronauts from spikes in radiation caused by solar storms is a major concern for long-duration space flights. When the Sun sends a huge mass of material toward Earth (called a coronal mass ejection), scientists on Earth prepare any astronauts in orbit for increased radiation exposure. All EVAs are canceled and the astronauts are told to move to the most heavily shielded areas of the craft. Usually the water tanks onboard provide the best shielding. No standing around the water cooler—try sitting behind the water tanks.

## DON'T FORGET YOUR TOWEL

Must-haves in space include towels and Velcro. Towels are essential for soaking up any loose liquid that's flying around. And if you don't want tools, pencils, clothes, paperwork, or anything else to float away from you, put Velcro on it.

## PRIVATE BEDROOM, NO VIEW

On the space station, there are six private bedrooms—well, they're more like little closets large enough to accommodate one person. Astronauts climb into a sleeping bag/harness that is fixed to the wall or floor, close the sliding door, and off they go to dreamland.

## THE SIXTY-TWO-MILE-HIGH CLUB

Has anyone joined the sixty-two-mile-high club and had sex in space? Officially NASA answers no. Some space experts agree, citing the fact that past astronauts had not enough time or privacy. Plus, gravity might not be your friend in the field of love. Every action has an equal and opposite reaction, so breaking up is easy to do.

## VELCRO IN AND ENJOY THE RIDE

Proponents for sex in space note that a married couple went into space (Jan Davis and Mark Lee), and Russian cosmonauts Yelena Kondakova and Valery Polyakov seemed to have their romance blossom in orbit, so who's to say what does and does not happen during those forty-five minutes of nighttime. And there is plenty of Velcro onboard to keep a relationship together.

## KEEP IT FRESH

If you want to shave in space, you can do so with an electric razor or with a regular razor and specially engineered NASA shaving cream. Just be sure to wipe the blade after each stroke with a towel to collect the hairs. And if you need a trim while visiting the ISS, use the onboard electric clippers, which are attached to a vacuum device.

## AFTER EVERY MEAL

To brush your teeth in space, you can either use edible toothpaste and swallow it afterword, or enjoy your favorite brand and spit it out in a towel when complete. Your crewmates will thank you.

## NO SHOWER, NO PROBLEM

Astronauts don't shower aboard the ISS. Instead they take sponge baths and use dry shampoo to wash their hair. Just to reiterate—moist sponges and dry shampoo are your main ways of washing your body over the course of a mission lasting several hundred days.

## DON'T USE ALL THE HOT WATER!

America's first space station, Skylab, actually had a shower. An astronaut could stand inside a cylindrical shower curtain, squirt water out of a jet, and let a system of vacuums suck the water away.

## I PUT THAT SAUCE ON EVERYTHING

When in space, the fluids in the body tend to settle in different places due to diminished gravity. Most astronauts experience clogged sinuses as if they had a head cold. Stuffed-up sinuses also make food in space taste much blander than on Earth. The best remedy: hot sauce. Astronauts have found that hot sauce helps on pretty much anything, and it has been a staple on space missions.

## ISS LOST AND FOUND

If you lose something onboard the ISS, check the vents. Miscellaneous items tend to gather at the air intake registers.

## ATOMIC BONDING

When two pieces of super-clean, similar metals touch in the vacuum of space, they stick together. Bonding occurs at an atomic level as one piece of metal can't tell where it stops and the other begins. Astronauts need to be really careful to prevent this "cold welding" from occurring when they're outside the spacecraft. Otherwise, there might be a lot of wrenches stuck to the side of the ISS.

## A DANGEROUS PROFESSION

Eighteen people have died on space missions: thirteen Americans, four Russians, and one Israeli. Only three people have died in space, while the remaining fifteen perished either shortly after liftoff or during the descent to Earth.

## ANGRY LAST WORDS

Soviet cosmonaut Vladimir Komarov was the first person to die during a space mission. During his nineteen orbits of Earth, his Soyuz 1 spacecraft had a string of mechanical failures. As he returned to Earth, Komarov took over manual control and maneuvered through the upper atmosphere. As he rapidly descended, the parachutes did not deploy, and he and his spacecraft struck the ground full force. During the descent, American listening posts in Turkey heard Komarov's lasts words over the airwaves, which consisted mainly of curses and cries of anger directed at the Soviet officials who sent him on this suicide mission.

## BURN, BABY, BURN

If you go into space without a suit, you will die very quickly. Lack of oxygen will knock you out within fifteen seconds—which is good, because you don't want to be aware of what's to come. Due to the low/no pressure in outer space, your fluids will boil away to such a degree that they will then freeze. Your organs will then rupture or, at the very least, stop working. If you were in low Earth orbit when this happened, you'd probably fall back to Earth. The friction of coming through the atmosphere would most likely burn up your body before it hit the ground.

## HAVE YOU GROWN?

Being in space makes you taller. When astronauts return to Earth from long periods of weightlessness, they may find that they have grown up to 3 percent taller since they last left Earth.

## TIME TO HIT THE GYM

As you read this book you are using more muscles than an astronaut on the treadmill aboard the ISS. The gravity on Earth works your body more than you might imagine. Astronauts in the microgravity of space need to exercise two hours per day in order to not be totally incapacitated when they return to Earth. Nevertheless, when astronauts return from long-duration flights, many cannot stand up upon landing.

## I DROPPED IT AGAIN

Some astronauts that have been in space for several months experience a weird phenomenon when they return to Earth: They forget about gravity. Most notably, they drop items, thinking that they'll just float next to them.

## SMALL SPACE LIVING

The record for the longest single space flight goes to Russian Valery Polyakov, who spent more than fourteen straight months aboard the Mir space station. That's over one year in the same seven modules, each the size of a small apartment. But with no up or down, he could use every inch of floor-to-ceiling space!

## ADDICTED TO SPACE

Russian cosmonaut Gennady Padalka has spent more time in space than any other human. He resided a total of 879 days in orbit over the course of five missions and flew on both the Mir and ISS.

## DO YOU THINK THEY'RE GEMINIS?

Over the course of four missions, American astronaut Scott Kelly spent more than one year in space. NASA is studying the effects of long-term flights like this on the human body. Scott's twin brother Mark, who remained on Earth during his brother's last 340-day stint on the ISS, is the control subject. After spending so much time in microgravity, Scott returned to Earth two inches taller than he was when he first blasted off into space.

# CHAPTER 2

# FLY ME TO THE MOON

## *Facts about Earth's Nearest Neighbor*

Circling a newborn star, a young planet has just formed. Still molten from the heat and pressure of its swirling cocoon of gases and heavy elements, the planet is continuously bombarded by stray meteors and asteroids. Then from deep space, another massive object comes into view. The object slams into the undeveloped planet with such tremendous force that chunks of it are ripped free and explode into space.

This is not the plot of the latest sci-fi novel. The planet is Earth 4.5 billion years ago. And the catastrophic incident describes the origin of the Moon.

The Moon is made of Earth's mantle that was blown off by a Mars-sized impactor slamming into the newly formed planet. Roughly 1 percent of Earth flew into space that day, and the resulting debris formed into a ring around the planet. Over eons, gravity brought many of the shattered pieces back together. They heated up, formed into a sphere of rock, and became the Moon you love to see on a warm evening. In this chapter you'll take a closer look at this world created by that singular destructive act.

# MOON BASICS

## WON'T YOU BE MY NEIGHBOR

Other than the occasional passing asteroid and manmade satellites, the Moon is Earth's closest neighbor in space. Since it orbits Earth, the Moon is also referred to as Earth's natural satellite.

## THE CONSTANTLY MOVING MOON

The Moon is, on average, 238,900 miles from Earth. But the Moon's distance from our planet constantly changes. When the Moon is closest to Earth, at about 221,500 miles, astronomers call that perigee. When the Moon is farthest from Earth, at almost 252,500 miles away, that is apogee. You can't tell the difference from night to night, but if you make a side-by-side comparison of a full moon at perigee versus a full moon at apogee, the difference is dramatic. A perigean full moon is over 31,000 miles closer and appears 14 percent larger in diameter and 30 percent larger in surface area than an apogean full moon.

## CLOSER THAN YOU THINK

The Moon's distance from Earth (roughly 238,900 miles) is based on the distance from the center of Earth to the center of the Moon. But the distance from the surface of Earth to the surface of the Moon is much closer. You live about 4,000 miles above the center of Earth. And the lunar surface rests more than 1,000 miles from its core. When you factor everything in, the closest you can be to the Moon and still be on the surface of Earth is about 218,000 miles.

## MODEL EXERCISE

If you want to picture the size and distance between Earth and the Moon, do this: Take a standard twelve-inch-diameter globe, the kind that is found in most elementary school classrooms. Then get a three-inch-diameter Styrofoam ball. Put that ball next to the globe and you have the relative sizes of Earth and the Moon. To demonstrate their relative distance, place the Styrofoam ball thirty Earth diameters (thirty feet) away.

## THE SUPERMOON

The closest full moon in a calendar year is known as a supermoon. This is a term that only became popular in 2012 when the media picked up on this astronomical fluctuation in distance. Astronomers refer to this event by a less catchy name: perigee-syzygy. To date, several names for the farthest full moon of the year have been proposed, including puny moon, tiny moon, and micro moon. The media has not settled on which name they prefer.

## DON'T PUSH ME AWAY!

The Moon used to be closer to Earth. It moves about one inch farther from Earth every year. What's happening is that the Moon's gravity is sapping energy from the rotating Earth, and that energy pushes the Moon into a higher orbit. This also slows down the Moon's velocity around Earth and makes its orbit take a fraction of a second longer to circle Earth each year.

## PUTTING ON THE BRAKES

The Moon is responsible for slowing down Earth's rotation. This act is called tidal friction, and the Moon's gravity acts like brakes that slow down Earth's daily spin. Currently Earth takes twenty-four hours to rotate once. But each year it takes 15 millionths of a second longer for Earth to spin one time. It's not much, but it will add up over billions of years.

## SIZING UP THE MOON

The Moon is 2,159 miles in diameter (about the distance from Atlantic City, New Jersey, to Las Vegas, Nevada). If you put the Moon next to Earth, it is only a little more than $\frac{1}{4}$ the diameter of Earth.

## STEPPING ON THE SCALE

By volume the Moon is $\frac{1}{49}$ the size of Earth. However, it is much less dense than our planet, so if you put the two bodies on a scale, Earth would weigh eighty-one times more than the Moon.

## DOES SIZE MATTER?

Our Moon is the fifth-largest moon in the solar system. Only Jupiter's moons Ganymede, Callisto, and Io and Saturn's moon Titan are larger than the Moon.

## SUSPICIOUSLY LARGE MOON

The Moon and Earth have the largest moon-to-planet-size ratio in the solar system. While the Moon is more than $\frac{1}{4}$ the diameter of Earth, Jupiter's moon Ganymede—the largest moon in the solar system—is less than $\frac{1}{26}$ the diameter of giant Jupiter. That means that Earth, for its size, has an oddly large moon.

## PLANETARY BLASTS

The relatively large size of Earth's moon tipped off scientists that there was a chance the two objects were once one body. And, in fact, when Apollo astronauts brought Moon rocks back to Earth, geologists found evidence that those rocks shared a common history with Earth, which gave them physical proof that Moon rocks were blasted there by a giant collision with our planet.

## MOVING COUNTERCLOCKWISE

If you picture the North Pole as the top of Earth (i.e., north is up), Earth rotates in a counterclockwise direction. The Moon likewise revolves—or orbits—counterclockwise around Earth.

## A MOON BY ANY OTHER NAME

The Moon's name is a holdover from the word *mona*, an Old English word for the Moon. However, the Moon goes by different names in other cultures. To the ancient Greeks it was *selene*, in Latin and Spanish it is *luna*, and in Swahili it is *mwezi*.

# MOON PHASES

## LUNACY!

It's not lunacy. Moon experts can look at a picture of the moon and, based on its orientation, phase, and location in the sky, can tell you approximately what time of night and what day of the year the photo was taken.

## REFLECTING ON THE SUN

The Moon generates no light of its own—it shines with reflected sunlight. Light travels 93 million miles from the Sun, bounces off the surface of the Moon, and then travels another 240,000 miles to reach your eyes.

## WAX ON, WAX OFF

The Moon's phases are called new moon (when it is all dark), waxing crescent (a crescent with the right side illuminated), first quarter (when the right half is lit up), waxing gibbous (almost full with more light on the right side), full moon, waning gibbous (just past full with more light on the left side), third quarter (when the left half is lit), and waning crescent (when it looks like the letter C).

## GET A MOONDIAL

You can use the Moon's phases to tell time. Each phase is only visible at certain times of day. From the Northern Hemisphere, there is a saying to help: When the light is on the right, you see the Moon at night. When the left side shines, it's morning time. You will never see a waxing moon (light on the right) at sunrise, and you'll never see a waning moon (light on the left) around sunset.

## MOON PHASE MISCONCEPTIONS

Neither the shadow of Earth nor clouds cause the Moon's monthly phases (the top two misconceptions). The phases you see depend on where the Moon happens to be in its orbit around Earth. The Sun can only light up half of the Moon at one time. When you can't see this lit half, it's called a new moon. When the Moon moves ¼ of its orbit around Earth, you can see half of the lit half and half of the dark half from Earth, and it's now first quarter.

## SUNRISE, SUNSET

Every time a full moon rises, the Sun sets on the opposite side of the sky. Picture this: as you drive home from work and the Sun is going down behind you in the west, you see a big full moon rising above the eastern horizon. The full moon will also set just as the Sun rises.

## THE MOON'S MONTHLY CYCLES

It takes 27.3 days for the Moon to orbit Earth. But it takes about 29.5 days for it to go through one complete cycle of phases. The 27.3-day period is called a sidereal month and is the time it takes for the Moon to go around Earth and appear in front of the same background stars. During the 27.3 days, Earth has moved a significant distance around the Sun and so it takes the Moon an extra 2.2 days to show the same phase that it did 29.5 days prior. That's called a synodic month.

## NEW MOON OR NO MOON?

You can rarely, very rarely, see a new moon. A new moon occurs when the Moon is between Earth and the Sun. During this time, none of the sun's light reaches the side of the Moon that you can see. *And* the new moon is up in the sky the same time as the Sun. It's really difficult to see an object that is not receiving any light, especially when the much brighter Sun is also in the sky.

## THE SOLAR ECLIPSE EXCEPTION

The only time you can see a new moon is during a solar eclipse. During these rare events you can (with special protective eclipse glasses) see the new moon in front of the Sun. It is completely black, but it's there.

## WHO KNEW, BLUE MOON?

The definitions for *blue moon* are pretty sketchy. Some say that a blue moon is the name of the second full moon in a calendar month. For instance, you can have a full moon on January 1 and another full moon on January 30. This definition first appeared in a 1946 article in *Sky & Telescope* magazine. Or some say a blue moon is the third full moon in a season that has four full moons in it. Either way, astronomers never use this term.

## ONCE IN A BLUE MOON

The Moon can appear blue on rare occasions (thus the phrase "once in a blue moon"). This effect is caused by particles in Earth's atmosphere turning moonlight a shade bluer. Volcanic eruptions on Earth that eject dust and ash high into the air are generally the culprit of actual blue moons.

## A REVERSAL OF PHASES

If you visited the Moon, you would see Earth phases. You would see a new earth, crescent earth, gibbous earth, and full earth over the course of 29.5 days. These Earth phases would be completely opposite the Moon phases seen from Earth. So when you see a full moon, visitors on the Moon see a new earth.

## SUN SIZE

From the Moon, Earth looks four times wider and sixteen times larger than the Moon does from Earth. And if you were on the near side of the Moon, except during a "new earth," you would always see Earth in the sky. The Sun, however, would appear to be the same size on the Moon as it is on Earth.

## THAT'S NOT MOONSHINE

Have you ever looked at a crescent moon and also seen the dark part looking like a faintly glowing, gray ball? This only happens when the Moon appears as a slim crescent. This effect is caused by earthshine, which is the light of the Sun that shines on Earth and then reflects up on the Moon. Since the Moon is showing a crescent, the sunlight bouncing off a nearly full earth brightens the Moon enough so you can see it against the darkness of the sky.

## THE MOON ILLUSION

Have you noticed that the Moon looks bigger when it is near the horizon? This effect is called the Moon illusion. Some people believe that the Moon appears larger on the horizon because you have reference points like trees and buildings. Not true. The Moon illusion is caused by how you perceive the sky. You picture something on the horizon as farther away than something overhead in the sky. Since the Moon is actually the same size, the "farther" Moon on the horizon becomes larger in your mind. The illusion takes place in your brain. You can combat the Moon illusion by looking at it upside down, between your legs. For some reason the Moon will look normal-sized.

## SHINE ON, HARVEST MOON

The harvest moon is traditionally the full moon closest to the autumnal equinox (September 22 or 23), but it is only rarely the biggest, brightest full moon of the year. In the past, the harvest moon coincided with harvest time, and your farmer ancestors used its light to help harvest the crops by day or night.

## WHAT'S YOUR NAME, AGAIN?

Many cultures have named the full moons that occurred each month. According to the *Farmers' Almanac*: January = Wolf Moon, February = Snow Moon, March = Worm Moon, April = Pink Moon, May = Flower Moon, June = Strawberry Moon, July = Buck Moon, August = Sturgeon Moon, September = Harvest Moon (though it sometimes can be in October), October = Hunter's Moon, November = Beaver Moon, and December = Cold Moon.

## AN ANGELIC MOON

Have you even seen a ring around the Moon? It is called a lunar halo and it occurs when moonlight bends through lots of six-sided ice crystals in Earth's atmosphere creating a huge, perfect circle of light. The radius of a lunar halo is always a shade less than 22 degrees. If you extend your arm and spread your thumb and fingers, the span between your thumb and pinkie is about 25 degrees. The next time you see a lunar halo, stretch out your arm and cover the Moon with your thumb. Your pinkie should reach the edge of the halo.

## AN EPHEMERAL MOONBOW

Everyone has marveled at a rainbow, but is it possible to see a moonbow? You need a nearly full moon and a source of mist to interact with the moonlight. The best places in the United States to see these ghostly, nighttime glows of light are at Yosemite Falls in California and Cumberland Falls in Kentucky.

# ECLIPSES

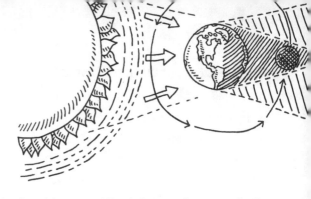

### LUCKY SEVEN

Eclipses happen when the Sun, Moon, and Earth line up almost perfectly. The complex cycle of these alignments means that from Earth you can see no more than seven eclipses in a calendar year. This last happened in 1982 when there were four partial solar eclipses and three total lunar eclipses.

### LUNAR ECLIPSE ALIGNMENT

A lunar eclipse occurs when Earth is between the Sun and Moon and casts a shadow over the lunar surface. That means that a lunar eclipse can only occur when the Moon is full.

### SHADOWS IN SPACE

Earth actually casts two shadows in space all the time. Even though the Sun is so far away, it's large enough to create a darker shadow right behind Earth, called the umbra, and a lighter shadow, called the penumbra. You can only see these shadows when they are cast onto things like the Moon and Earth.

### ON THE HORIZON

During sunset, if you look to the opposite side of the sky from the Sun, you will see a faint but distinct shadow just above the horizon. That is Earth's shadow cast onto its atmosphere. You can see that at sunrise as well.

## A SHIFTING SPECTRUM

During a total lunar eclipse, the Moon turns various shades of gray, orange, pink, and red. This is caused by rays of sunlight bending through Earth's atmosphere, which shift their color to the red side of the spectrum. This is the same reason that at sunset and sunrise the Sun looks red instead of its normal yellow color and the sky turns a shade of red. So when you see this color on an eclipsed Moon, you are actually seeing the sunlight going through the sunrises and sunsets of Earth and then reflecting off the Moon.

## BATHED IN RED

If you were standing on the Moon during a total lunar eclipse, you'd be bathed in red light. When you looked back at Earth, it would be all dark and would block out the Sun. From that lunar perspective you would see a solar eclipse!

## PARTIALLY SHADY

A partial lunar eclipse happens when the Sun, Earth, and Moon are not aligned precisely enough for the Moon to completely enter Earth's shadow. This looks just as it sounds: only part of the Moon is blocked out.

## A PRETTY LAME PENUMBRA

A penumbral lunar eclipse is when the Moon enters the penumbra, the lighter shadow of Earth. This shadow, cast on a full moon, is almost completely invisible to the average observer. So if you hear about a penumbral eclipse coming up, don't get too excited. You won't notice anything.

## SOLAR ECLIPSE CONFIGURATION

A solar eclipse occurs when the Moon stands between the Sun and Earth and casts a shadow onto Earth's surface. This can only happen during a new moon.

## A PAC-MAN SUN

A partial solar eclipse—when only part of the Sun is blocked by the Moon—is the most common type of solar eclipse to witness. Maybe you've seen pictures of crescent suns. They look just like Pac-Man.

## CRESCENT SUNS ALL AROUND

During strong partial solar eclipses (when 75–95 percent of the Sun is blocked by the Moon), the sunlight can play some cool tricks. If you look at how light goes through the leaves on trees and hits the ground during good partial eclipses, you'll see that the beams of light make crescent suns. The spaces between the leaves act like pinhole cameras, and instead of casting circular beams of light, they display multiple copies of the eclipse going on.

## HEAVENLY PRECISION

There is an amazing precision with total solar eclipses. Since the Moon can just barely block out the Sun, what you see depends on where you are on Earth. You have to be in the right place at the right time to experience it. The total solar eclipse path is generally less than 100 miles wide and can only be visible from about 1 percent of Earth's surface.

## JET-SET ECLIPSE

Unfortunately, total solar eclipses only last a short time. They can be as short as one second and as long as seven minutes and thirty-  two seconds. The Moon's shadow moves across Earth's surface faster than any jet, and although you can't keep up with an eclipse, people try. If you're flying in a plane through the Moon's shadow during a total solar eclipse, you can experience a few extra seconds of totality.

## SAYONARA, SOLAR ECLIPSES

Since the Moon is moving away from Earth, eventually it will be too far away from Earth to create total solar eclipses. The Moon's umbral shadow will no longer reach the surface of Earth. Fortunately, that won't happen for millions of years.

## DIAMONDS ARE AN ASTRONOMER'S BEST FRIEND

The moment before and the moment after a total solar eclipse you can see a ring of light around the Moon that astronomers call the diamond ring. You can also see bits of Sun come around the imperfect edges of the Moon in little balls of light, called Baily's beads. What you are seeing is the sunlight silhouetting the mountains and valleys on the rocky world above.

# THE MOON THROUGH A TELESCOPE

## I'LL BE BACK

Round craters, high mountains, and deep valleys are all on display whenever you look at the Moon through a small telescope. The line dividing the day from the night on the Moon is called the terminator. The terminator is the best place to look with a telescope because that's where you can see dramatic plays of light and shadow on the uneven lunar landscape.

## CRATER FACE!

There are so many craters on the Moon that they are hard to count. There are even craters inside craters. These were all made by asteroids, meteors, and comets slamming into the lunar surface a long time ago.

## NERDY AND ARTSY

Astronomers used to be artists. In the 1800s, astronomers had a difficult time photographing the Moon. The pictures almost never looked as good as the real thing. So artist/astronomers created 3D models of the lunar landscapes that they saw in their telescopes. Then they would take pictures of their models with the appropriate lighting and share them with the public and other scientists. These models were so detailed that you would have a hard time telling them from the real thing.

## MOON MOUNTAINS

The smallest Moon craters you can see through a backyard telescope are about the size of a city like San Francisco. Some of the larger craters are ringed by bright rays of materials that were ejected during the moment of impact when the crater was created. Some of these rays radiate out for hundreds of miles. In the center of some of the larger craters are tall, steep mountains that were created when materials fell back into the center after the huge impact that created them. When the sunlight strikes the craters at a low angle these peaks may glow while the crater floor is all dark, appearing as islands of light in a sea of darkness.

## LUNAR SEA LANDING

There are two main terrains on the Moon: the brighter areas, called the highlands, and the darker areas, called the seas, or maria. The most famous mare is the Sea of Tranquility where Neil Armstrong and Buzz Aldrin landed in 1969.

## CRACKING THE SURFACE OF THE MOON

Maria (the Moon's seas) are mostly round. They are the result of cataclysmic impacts from meteoroids that carved out huge swaths of the lunar terrain. Lava seeped up through cracks from the interior of the Moon and partially filled in the giant craters. The maria are now flat, hardened lava plains.

## THE FAR SIDE OF THE MOON

No matter the day, month, or year, you can only see one side of the Moon. The Moon is tidally locked to Earth, so unless you leave Earth you can never see the other side. Astronomers call that the far side of the Moon.

## DON'T FEAR THE DARK SIDE

The far side of the Moon and the dark side of the Moon are not the same thing. Whereas the far side is always the far side, the dark side is a fluid geographical term. The dark side of the Moon is the part of the Moon not currently illuminated by the Sun. The dark side of the Moon could also be the far side of the Moon during a full moon. But when there is a new moon on Earth, the dark side is also the near side.

## THE MOON IS TWO-FACED

The far side of the Moon looks quite different than the near side. There are greater numbers of smaller craters on the far side and much fewer maria. Both sides of the Moon probably received about the same amount of impacts, but lava flows that filled in the maria on the near side most likely covered over the older, smaller craters. The far side had no such crust-shattering impacts that created extensive lava flows and maria.

## IN SYNCH

Many people think that the Moon does not rotate, but this is incorrect. The Moon rotates one time for every one revolution it makes around Earth. It's called synchronous rotation and is very common among moons in the solar system. That's why you only see one face of the Moon from Earth. If the Moon did not spin, it would face the same direction in space and you'd see the other side of the Moon from time to time.

## LUNAR LIBRATION

You'd think you could only see 50 percent of the
Moon, but sometimes you can get glimpses
around the edges. When the Moon is a little
farther north than normal you can see under
it a bit. When the Moon is a little farther
south, you can see a little over the top.
And when the Moon is going a little faster
or slower you can see around the eastern or
western edges. This effect is called libration, and
keen Moon watchers since the seventeenth century
took advantage of these peeks into the far side to map an
extra 9 percent of the lunar surface even before spacecraft had photographed the far side.

# MISSIONS TO THE MOON

## MOON MANDATE

In 1961, President John F. Kennedy raised the standard for American space exploration and declared that the United States would put a man on the Moon. At 240,000 miles away, reaching the Moon required transporting a capsule 800 times farther than any previous mission. NASA's Apollo program, which employed half a million people and commanded billions of tax dollars, met this challenge.

## A WEEKLONG VOYAGE

It takes three days to get to the Moon and three days to return. NASA developed a new rocket in the late 1960s, the Saturn V, to take the Apollo astronauts to the Moon and back. The Saturn V stood 363 feet tall (equal or greater in height to many city skyscrapers) and remains the largest launch vehicle ever used.

## TEST TRAGEDY

The Apollo program began tragically. Apollo 1, manned by astronauts Gus Grissom, Ed White, and Roger Chaffee, didn't even get off the ground. During tests on a Florida launch pad, fire engulfed the capsule and killed all three crew members. After months of investigations and safety design changes, the Apollo program pushed ahead with a series of unmanned flights (Apollo missions 4, 5, and 6).

## A SPACECRAFT WITH A VIEW

On Christmas Eve 1968, astronauts Frank Borman, Jim Lovell, and William Anders, aboard Apollo 8, became the first people to orbit the Moon. When they first rounded the far side of the Moon, they beheld an amazing sight: their home planet. That night the crew sent a live broadcast back to Earth to share their thoughts and observations. Lovell, obviously moved, commented, "The vast loneliness is awe inspiring and it makes you realize just what you have back there on Earth." Anders took the now iconic picture showing an earthrise over the stark lunar landscape.

## NAMING RIGHTS

In 1959, the unmanned Soviet spacecraft Luna 3 captured the first images of the far side of the Moon. Since they were first to reach the mysterious, unseen realm, the Soviet Union named many of the lunar features over there.

## ALMOST THERE . . .

The Apollo 10 mission did everything but land on the Moon. The capsule commanded by Tom Stafford successfully made the journey to the Moon, descended to 47,000 feet off the lunar surface, and returned the crew safely. The stage was now set for Apollo 11's historic flight.

## THE *EAGLE* HAS LANDED

On July 20, 1969, while Michael Collins remained in orbit sixty miles above, Neil Armstrong and Buzz Aldrin descended to the Moon in the Lunar Module named **Eagle**. After a few tense moments while Armstrong took over manual control of the landing to avoid a boulder field, they eased onto the lunar soil. As millions held their breath on Earth, and with only thirty seconds of fuel remaining, Armstrong announced, "The **Eagle** has landed." Apollo 11 was a success!

## WHAT'D YOU SAY?

Buzz Aldrin uttered the first words transmitted from the surface of the Moon. Upon touchdown, he said, "Okay, engines stop," a moment before Armstrong said, "The **Eagle** has landed."

## A LIGHT LUNAR LANDING

Neil Armstrong's heart rate went from 75 to 150 beats per minute when he took over manual control of the Moon landing from the computer. He landed the **Eagle** so lightly that it barely sunk into the lunar soil. This made their ladder from the **Eagle** to the Moon too short to fully reach the ground. The astronauts would need to leap, not step, onto the Moon.

## MARKING THE OCCASION

Armstrong and Aldrin were eager to conduct their EVA (extravehicular activity). After having a bite to eat, Armstrong crawled out the hatch feet first, down the ladder, and onto the dusty lunar landscape. Slightly out of breath, he spoke those memorable and eloquent words, "That's one small step for a man, one giant leap for mankind."

## LUNAR LEAK

Aldrin followed closely behind and has another distinction: the first person to pee on the Moon. He urinated inside a container inside his suit. When he returned to the *Eagle* and jumped up to the first rung of the ladder, the container's seal was compromised. Urine leaked out and into his boot.

## A QUICK LAYOVER

When they were on the Moon's surface, Armstrong and Aldrin collected rocks and soil, took pictures, and set up scientific equipment for two hours. Compared to later Moon landings that lasted up to three days, Armstrong and Aldrin had a very brief visit. But apparently those two hours were exhausting! Upon entering the lander, the astronauts took a brief nap before launching from the Sea of Tranquility and rendezvousing with Collins to head for home.

## MOON MOOD MUSIC

Neil Armstrong brought music to listen to during his journey to the Moon and back. His mix tape included the *New World Symphony* by Antonín Dvořák and one song from the album *Music Out of the Moon: Music Unusual Featuring the Theremin*.

## HOUSTON, WE HAVE A PROBLEM

Apollo missions 12, 14, 15, 16, and 17 all made nearly flawless landings on the Moon, but you can't say the same about unlucky mission 13. Fifty-five hours into their flight on Apollo 13, Jim Lovell (on his second Apollo flight), Jack Swigert, and Fred Haise heard an explosion in one of the oxygen tanks that knocked out two of their three fuel cells. For four days NASA engineers worked feverishly with the crew to correct the problems, and the nation tuned into every news update. Thanks to the ingenuity of NASA's Mission Control and the resourcefulness of the Apollo 13 crew, the astronauts splashed down safely into the South Pacific. Although Apollo 13 failed its mission, it became the most inspiring failure in the history of space flight.

## A LONG DRIVE

Astronaut Alan Shepard smuggled a golf club and golf balls aboard his Apollo 14 flight to the Moon. While on the lunar surface he surprised NASA by pulling out a six-iron and hitting a few balls. His spacesuit was too stiff for him to use both hands, but after two mediocre chips, the third ball traveled, according to the golfer himself, "miles and miles."

## BYE-BYE BUGGIES!

Apollo missions 15, 16, and 17 made the lunar rover—or Moon buggy—famous. Despite the reduced gravity, astronauts had found it difficult to travel much farther than a few hundred feet on the Moon due to their cumbersome spacesuits. Lunar rovers greatly increased the range of their EVAs. The rovers were electric and could cruise at the speed of 8 miles per hour (although at one point Gene Cernan topped 11 miles per hour during his lunar rover drive on the Apollo 17 mission). All three lunar rovers were left on the Moon to reduce the weight of the spacecraft and make room for additional Moon rocks.

## LUNAR LITTERBUGS

Humans have left more than 400,000 pounds of trash on the Moon. From golf balls and bags of pee to lunar rovers and crashed spacecraft, NASA has amassed a twenty-two-page list of every item that remains on the Moon. In exchange for what they left behind, Apollo astronauts brought 838 pounds of Moon rocks back to Earth.

## SUPERMAN'S STRENGTH

The downward pull of gravity on the Moon is a mere 1/6 of what it is down here on Earth. Imagine that: instead of weighing 180 pounds, you'd weigh 30, but you'd have the same muscles that you have on Earth. So when you get to the Moon, you're like Superman. You could jump higher and lift formerly heavy things. Heck, if the lunar rover gets stuck, you could just pick it up and carry it for a while!

## ESCAPE FROM THE MOON

Can you jump off the Moon? Not unless you can reach escape velocity (the speed required to fully break free of the gravity of an object), which is more than 5,000 miles per hour on the Moon. Although the gravity on the Moon is much less than that on Earth, you'd still need a lot of power to leave it.

## ROCKET MEN

Twenty-four men have flown around the Moon. Twelve of those astronauts also walked on the Moon, including Neil Armstrong, Buzz Aldrin, Pete Conrad, Alan Bean, Alan Shepard, Edgar Mitchell, David Scott, James Irwin, John Young, Charles Duke, Gene Cernan, and Harrison Schmitt.

## IT DOESN'T SMELL LIKE CHEESE?

Many astronauts have commented on the distinctive smell of moon dust. Some liken it to spent gunpowder or wet ashes from a fireplace. Apollo 16 astronaut John Young claimed that it didn't taste "half bad."

## THE LAST VISITORS TO THE MOON

The final mission to the Moon was Apollo 17. On this mission, astronauts Gene Cernan and Harrison Schmitt travelled 22 miles with the lunar rover, collected 243 pounds of rocks, and took 2,000 pictures. On December 14, 1972, Cernan and Schmitt left the Moon to return home and no one has returned since.

## A LAST WORD, PLEASE

Cernan believed that the last words he said on the Moon were, "Let's get this mother out of here." He may have said those words, but they were not recorded in the NASA transcript. At one point, Cernan said, "Okay now, let's get off. Forget the camera," but fourteen seconds later, just before liftoff, Harrison has the honor of saying the final words from the Moon when he signals, "Ignition."

## A CLOSE-UP LOOK AT LANDING SITES

Although humans have not returned to the surface of the Moon, NASA has kept a close eye on it since 2009 thanks to the unmanned craft called the Lunar Reconnaissance Orbiter (LRO). LRO has captured some of the most high-resolution images of the lunar surface to date, and during several flybys it even imaged the landing spots of the Apollo missions. In some pictures you can see the landers, instruments, and even well-worn paths that astronauts carved with their footsteps on the lunar soil.

## ROVING RECORD

The unmanned Soviet lunar rover Lunokhod 2 travelled twenty-four miles on the Moon during its four months of operation in 1973. This eight-wheeled rover held the off-Earth roving record until the Mars rover Opportunity topped it in 2014.

## BACK TO THE MOON

NASA is currently debating where to send humans in the next two decades. Should they return to the Moon, go on to Mars, or try somewhere in between? Why return to the Moon? NASA has been there, done that. True. But they haven't done that in over forty years and could use the practice. If something goes wrong, it would be much better if it happened only 240,000 miles, or 1.3 light seconds, away. Fixes, rescues, and communication can be done in a timely fashion. Going to the Moon could be a practice run for farther and better things.

# CHAPTER 3

# INSIDE AND OUTSIDE THE SUN

*All about the Star Attraction*

Throughout the history of mankind, the Sun has ingrained a daily rhythm into the very nature of life on Earth. The repetition of the Sun rising, setting, and rising again became the primary cycle of life on Earth. Call that a "day." The first rays of light at sunrise made a welcome sight and led to a day of hunting, gathering, and farming. Ancient observers monitored the way light fell at different times of the day and how cast shadows moved, shortened, and lengthened. As the Sun set, they noted the emergence of different animals and wondered how long it would be before the Sun would return. Eventually, they comprehended the Sun's longer pattern—call that a "year"—and from then on, every day of every year, the first astronomers could predict when and where the morning Sun would rise and the evening Sun would set.

The Sun taught early sky watchers to be scientists, and in turn, modern astronomers have learned a lot about Earth's solar neighbor. Here you'll learn everything you need to know about the Sun: from its sweltering core to its extensive reach in the solar system.

# SOLAR BASICS

### LONE STAR

The Sun is a star, and it is the only star in the solar system. It contains 99 percent of the mass of the entire solar system and is the source of almost all the light and heat for the planets, moons, and asteroids. It is the focal point of the solar system.

### 109 EARTHS WIDE

The Sun's width, or diameter, is about 865,000 miles. When you compare that to Earth's diameter of 7,920 miles, you could fit 109 Earths end to end across the breadth of the Sun.

### LARGER THAN LIFE

By sheer size (by volume) the Sun dwarfs all the planets put together. It is 987 times larger than Jupiter, 22,500 times larger than Neptune, 1,300,000 times larger than Earth, and 200,000,000 times larger than Pluto.

### ALMOST PERFECT

The Sun is the most spherical object in the solar system. Unlike the planets, which are flattened at the poles, and smaller objects like moons, asteroids, and comets that are irregularly shaped, the Sun is almost a perfect sphere.

### A MASS OF GAS

The Sun is made almost entirely of gases. Hydrogen and helium, respectively, make up 75 percent and 25 percent of the Sun's mass. That doesn't leave much room for any other elements like carbon, nitrogen, and oxygen, which are there, but in very small proportions.

## A FORCE TO BE RECKONED WITH

Despite being made of gases, the Sun has mass and therefore gravity. Gas has mass, and a whole lot of gas can make enough gravity to not only hold the Sun together but to keep the planets circling around it instead of flying off into deep space.

## REACH OF THE STARS

The gravitational reach of the Sun is a sphere of space that ranges about 11 trillion miles in diameter—or 2 light years. Comets can travel that far from the Sun and its far-reaching gravity can still keep them within the solar system.

## NOT THE CENTER OF THE UNIVERSE

The Sun is not in the center of the Milky Way galaxy. Along with a host of other stars, the Sun resides in a region called the Orion Arm of the Milky Way, which lies closer to the edge of the galaxy than the center.

## THE SUN TAKES CENTER STAGE

Everything in the solar system orbits the Sun. This includes 8 planets; 5 dwarf planets; more than 600,000 asteroids; more than 1,000 plutoids (objects in the neighborhood of Pluto's orbit); more than 170 moons; numerous Kuiper belt and scattered-disc objects, as well as other classes of objects called centaurs, damocloids, trojans, cubewanos, twotinos, and plutinos; countless comets; and you.

## SLOW AND STEADY SUN

Moons revolve around planets. Planets revolve around the Sun. What does the Sun revolve around? The galaxy, of course. As one star in several hundred billion stars in the Milky Way, the Sun revolves around the center of mass, the center of the galaxy, where most of its stars live. It takes the Sun about 250 million Earth years to complete one orbit of the Milky Way. So in the Sun's almost 5-billion-year lifespan, it's completed about twenty trips around the galaxy.

## SPIN CLASS

The Sun also spins. It rotates counterclockwise, not as a solid body, but as a huge sphere of gases. This means that different parts of the Sun spin at different rates. At the Sun's equator, it takes about twenty-six days for the solar material to rotate one time. The Sun turns more slowly at the poles, where it takes about thirty-five days to complete one spin.

## GALILEO SUN STORY

In the early 1600s, Italian astronomer Galileo Galilei used his telescope to project an image of the Sun onto paper. With this projection method, he traced sunspots and charted their movements from day to day. These daily drawings gave him information that told him how the Sun rotated. With this, Galileo estimated the period of the Sun's rotation to be around twenty-eight days—very close to the actual period.

## SURPRISINGLY STEADY

The Sun is a combination of volatile forces and consistency. Continually turning itself inside out, the Sun erupts millions of tons of material into the solar system daily. However, the Sun's outpouring of energy is amazingly steady. Luckily for us Earthlings, the Sun's size and temperature does not fluctuate wildly like other observable stars, or Earth's weather would be even more unpredictable.

## HALFWAY THERE

The Sun started shining about 5 billion years ago. And astronomers predict that the Sun will maintain its consistent nature for another 4–5 billion years.

# LAYERS OF THE SUN

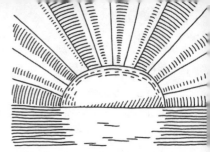

## ONE HOT NUMBER

What's the temperature of the Sun? It depends on where you measure it. Astronomers estimate that the core of the Sun is about 27,000,000°F.

## THE SUN'S SURFACE

Astronomers found that a star's color can also indicate its temperature. The bright yellow surface of the Sun, called the photosphere, is roughly 10,000°F.

## HOT SPOTS

Sunspots are a common feature on the photosphere and look like, well, spots on the Sun. They are the sites of magnetic storms—places where solar material is either shooting up off the sun's surface or falling back down in arcs and loops.

## COOLER BUT NOT COLD

While sunspots are still unbelievably hot and bright, they appear darker than the surrounding areas because they're cooler by about 2,000°F. The darker sunspots measure in at 8,000°F compared to their 10,000°F surroundings.

## DON'T TRY THIS AT HOME

Chinese astronomers were the first to document sunspots and observed especially large sunspot groups about 2,000 years ago with the naked eye. They did not employ safe, modern filters to observe the Sun but perhaps saw sunspots during partly cloudy days, through smoked glass, or reflected off still ponds. Please do not try these methods yourself. They will damage your eyes!

## UNSAFE FOR THE SUN

Never use the following as solar filters: tinfoil, foil wrappers, CDs, DVDs, x-rays, film negatives, smoked glass, regular sunglasses (even with UV protection), or Mylar balloons. In the past, many media outlets have claimed these are safe, but don't buy it. They will not adequately protect you.

## SAFETY FIRST

If you want to look at large sunspots today, the easiest and cheapest way to observe the Sun safely is by looking through a dark pane of #14 welder's glass, which costs around $5. Specially made solar eclipse glasses, called Eclipse Shades, use a black polymer and produce the same optical effects as #14 welder's glass for $1–$2 a pair.

## THEY'RE UNPREDICTABLE!

Predicting sunspots is really tough. Astronomers do not know when they will crop up, how large they will expand, and how long they will last. Some sunspots last a matter of hours. Others can last more than a month. That said, for the last 400 years astronomers have been keeping tabs on sunspot activity, and they've found a rough cycle of solar activity that repeats itself about every eleven years. Roughly every eleven years, there are periods of solar maximum when there are the most sunspots and periods of solar minimum when there is hardly any activity. But from day to day, there's no telling how many sunspots will be active.

## RIDE OUT THE STORM

Some sunspots can take an entire spin around the Sun. If they remain active for more than one month, they will be carried around by the rotation of the Sun, go around the backside of the Sun, and come around again to the front.

## THE COOLEST CHROMOSPHERE

Above the photosphere is the next distinct, gaseous layer of the Sun, called the chromosphere. The chromosphere is the part of the Sun where its gases have cooled off to a mere 7,800°F. You can see the chromosphere during a total solar eclipse.

## COOLER, COOLER, COOLER, HOTTER?

Here's a mystery: The core is the hottest part of the Sun, and as you move farther out from the core the temperature naturally drops. So the photosphere is cooler than the core and the chromosphere is cooler than the photosphere. But the temperature of the corona, the layer above the chromosphere, jumps up to as high as 1,000,000°F–3,000,000°F! Why? Astronomers don't know yet.

## UP IN THE ATMOSPHERE

Where the Sun ends is kind of nebulous. The Sun is a ball of gas, but its gases are present throughout the solar system. In fact, Earth, since it is bombarded daily by the Sun's solar wind, technically resides inside the Sun's atmosphere.

## SUPER INTERSTELLAR SPACE

The heliopause is where the solar wind peters out and mingles with interstellar space. The Voyager 1 spacecraft, launched in 1977, entered this region almost 12 billion miles from the Sun in 2012.

# SOLAR EXPLOSIONS

## AN EXPLOSIVE PERSONALITY

The Sun has an explosive personality. Although it seems to be a steady sphere of light, the Sun is continually in flux, erupting randomly and in all directions. The eruptions are called solar prominences, solar flares, and coronal mass ejections. Even the mildest outbursts from the Sun could engulf the entire Earth.

## PROMINENT ERUPTIONS

Along the edge of the Sun, astronomers can detect eruptions of stellar material that shoot out thousands of miles from the surface. One common and beautiful type of eruption is called a solar prominence. With a solar prominence, material spirals off the surface of the Sun in dramatic loops, arcs, and coils. Once the material cools, it falls back to the surface of the Sun. The shape of a solar prominence can change in a matter of minutes.

## FIERY FLARE-UPS

Solar flares are violent solar outbursts. Material wells up over a magnetic disturbance and creates a bubble-like formation. When enough pressure builds up, the flare bursts forth like an arrow leaving a bow.

## CORONAL MASS EJECTIONS

The most powerful explosions on the Sun are called coronal mass ejections, or CMEs. CMEs are so powerful that material breaks free of the Sun's gravitational pull and rockets into space at more than 1 million miles per hour.

## SOLAR LIGHT SHOW

Occasionally a coronal mass ejection barrels right toward Earth, causing solar particles to slam into the planet. Fortunately, Earth is protected by an invisible magnetic field that deflects the vast majority of these particles. However, some of this material reaches the upper atmosphere and becomes trapped. When these particles collide with one another they create amazing auroras. This typically happens around the polar regions and creates the northern and southern lights, but a particularly powerful CME can cause auroras to be visible in the northern and central parts of the United States.

## INCOMING!

Even though coronal mass ejections travel at more than 1 million miles per hour, they don't run into Earth unannounced. The Sun is 93 million miles away, so even at that high velocity, the solar material still takes two to four days to reach Earth.

## CAN YOU HEAR ME NOW?

Solar storms may have an effect on Earth. Although there is no direct evidence linking solar activity and global climate as of yet, what happens on the Sun 93,000,000 miles away can affect technology on Earth. Coronal mass ejections have been known to damage satellites, interrupt communication systems, and damage power grids. If your phone loses service, the Sun may be to blame.

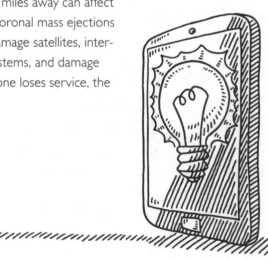

# THE SUN FROM EARTH

## NO PERFECT ORBIT

On average, the Sun is about 93 million miles away, but Earth's orbit around the Sun is not a perfect circle. Sometimes it is a little closer to the Sun than at other times. Each year around January 4 Earth is at perihelion, the closest point to the Sun, at 91.4 million miles away. Earth reaches aphelion, the farthest distance from the Sun, at about 94.5 million miles away, annually around July 4. Since there is only a 3 percent difference between perihelion and aphelion, the varying closeness to the Sun does not cause the seasons.

## THE REASON FOR THE SEASONS

The tilt of Earth causes the seasons. When Earth is tilted toward the Sun, the Sun rides higher in the sky, provides more direct energy, and offers extra hours of daylight. When Earth is tilted away from the Sun, the Sun crawls lower across the sky, Earth receives less energy per square inch, and there are fewer hours of daylight. This is what makes summer so much different than winter.

## THE LONG AND THE SHORT OF IT

The longest day and shortest nights of the year are well celebrated around the globe. June 21 is the summer solstice in the Northern Hemisphere of Earth where it is tilted most directly toward the Sun and soaks up the most solar energy. The Sun rises and sets farther north on the summer solstice than on any other day, and this event places the sun almost directly overhead at local noon for the mainland of the United States. After the summer solstice the Sun begins its slow trip south until the Northern Hemisphere experiences its longest night of the year, the winter solstice, on December 21.

## KEEP STILL!

The word *solstice* means "Sun standing still." This term was coined by ancient Sun watchers who noticed that the Sun rose and set in different places each day. However, on the solstices (near June 21 and December 21), the Sun seems to rise in the same place (and set in the same place) for several days in a row. The Sun itself does not stand still, but the positions of the sunrises and sunsets do on those days.

## A PERFECT BALANCE

On the spring (vernal) and fall (autumnal) equinoxes, everywhere on Earth receives twelve hours of daylight and twelve hours of darkness. Equal day and equal night equals equinox.

## IT MAY SEEM LIKE IT'S STRAIGHT OVERHEAD

Other than in Hawaii, the Sun never gets perfectly straight overhead in the United States or Canada. In Cincinnati, for example, the Sun's maximum elevation above the horizon is 74.5 degrees. The farther south you travel (in the United States or Canada), the higher the Sun appears. This is the reason people take vacations to Florida in the winter, to get more energy from the higher-arcing Sun.

## SUDDEN SUNSET

On the equator the Sun rises straight up in the sky and sets straight down. That's where you'd see the fastest sunrises and sunsets on Earth.

## BETWEEN THE TROPICS

The Tropic of Cancer marks the latitude of Earth where the Sun is perfectly overhead (known as a zenith) at noon on June 21, the summer solstice. If you're on the Tropic of Capricorn, the Sun does this on December 21, the winter solstice. Every point between the Tropics (also known as the Torrid Zone) experiences the Sun at the zenith at least once throughout the year.

## 23.5 DEGREES FROM THE POLE

What defines the Arctic and Antarctic Circles? It's astronomical! They lie at the latitudes 66.5 degrees north and 66.5 degrees south, respectively. That's 23.5 degrees from 90 degrees. So on June 21 when the Northern Hemisphere is tilted toward the Sun, everyone inside the Arctic Circle receives twenty-four hours of daylight. The Sun never sets. Everyone inside the Antarctic Circle that day experiences twenty-four hours of darkness. The Sun never rises. And the roles reverse come December 21.

## DELAYED SUNSET

The Sun is eight light minutes away. Sunlight takes a little over eight minutes to travel from the Sun to your eyes. That means when you watch a beautiful sunset, in reality the Sun already set eight minutes ago.

## WALKING ON THE SUN

If you're standing on the Sun, the brightest planet is Venus. Although Mercury is much closer to the Sun than Venus, Earth's sister planet is so much larger and reflects so much more sunlight that it would outshine the closer planet. Mercury would be the second-brightest while Earth would be third and Mars would be fourth.

## DON'T COME BETWEEN US!

Only three natural objects in space can predictably come between Earth and the Sun: the Moon, Mercury, and Venus. When the Moon blocks out the Sun it is called a solar eclipse. When either Mercury or Venus obscures a tiny portion of the solar disc it is called a transit.

## NOT SUCH A SHINING STAR

If you looked back at the Sun from a distant star system, it would not stand out. Say you were 24 light years away on the planet orbiting the star Fomalhaut: What would the Sun look like? It would be a fourth-magnitude star, much dimmer than any star in the Big Dipper. In fact, you'd be hard-pressed to even notice it.

# MISSIONS TO THE SUN

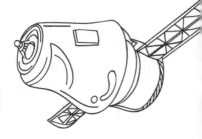

## KEEP AN EYE OUT

One of the more unsung, wildly successful, and durable missions is the Solar and Heliospheric Observatory (SOHO). This joint venture between NASA and the European Space Agency (ESA) has been watching the Sun almost every day since its launch in 1995. SOHO has documented many changes to the Sun over its two decades in space.

## GO THE LONG WAY

The Ulysses space probe, launched in 1990, took a circuitous route to view the Sun. It first flew away from the Sun and went out to Jupiter. Jupiter's massive gravity helped propel the spacecraft into an orbit that took it over the north and south poles of the Sun.

## SEE THE SUN IN STEREO

In 2006, NASA launched two separate rockets with one unique mission: to observe the entirety of the Sun. These twin Solar Terrestrial Relations Observatory (STEREO) spacecraft left Earth in opposite directions in order to flank the Sun. In 2011, these spacecraft were on exact opposite sides of the Sun. For the first time in human history, astronomers could see what was happening on not only the side of the Sun facing Earth but on the far side of the Sun as well, and they found that pretty much the same thing happens on both sides of the sun. That said, the resulting images are perfect for viewing with 3D glasses. You can visit STEREO's galleries at: http://stereo.gsfc.nasa.gov/gallery/gallery.shtml and see the Sun in three dimensions.

## CONTINUOUS SURVEILLANCE

The Solar Dynamics Observatory, launched in 2010, observes the ferociousness of the solar surface. The spacecraft takes high-resolution images of the Sun every ten seconds in ten different wavelengths of light. NASA scientists then stitch the images together to make awesome movies out of a week's worth of solar activity. With this information, they can try to discover the causes of solar outbursts and measure any effects to Earth.

## CRASHING TO EARTH

The Genesis mission attempted to collect samples of solar material and return them back to Earth. Scientists would then possess actual pieces of the Sun! The collection part went great and the return journey went very well until Genesis entered Earth's atmosphere and failed to open its parachute. The craft plummeted to Earth and crashed in the Utah desert. Despite the crash landing some solar samples remained intact and pristine.

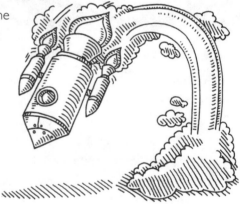

# THE SUN: PAST AND FUTURE

### YOU'RE A STAR!

The Sun formed from a huge, swirling cloud of gas and dust called a nebula almost 5 billion years ago. As the gases condensed and the nebula collapsed, the temperature in its center rose steadily. Eventually the temperature and pressure became so great that hydrogen began to fuse into helium. That's when the Sun was born. It "turned on" and began to emit massive quantities of heat and light.

### THE SOLAR WIND

The outpouring of energy from the newborn star dispersed the leftover gases in the nebula. This solar wind is still present today and streams particles from the Sun throughout the solar system.

### EVERYONE LOVES LEFTOVERS

After the Sun was born, the planets formed from the leftover parts that swirled around the newborn star. As eons passed, rocky material clumped together to form the inner planets (Mercury, Venus, Earth, and Mars), and gases amassed to make the outer planets (Jupiter, Saturn, Uranus, and Neptune).

### THE ORIGINAL SPIN

The counterclockwise spin seen in the Sun today (where it rotates about once a month) is the same spin that the nebula that formed the Sun had. In fact, all the planets in the solar system orbit the Sun in that same direction, and almost all the planets rotate counterclockwise as well.

## EVER-EXPANDING GASES

The Sun has gotten larger since it first formed. That may sound confusing because it's continuously consuming its hydrogen gas as fuel—which you would think would make it smaller, but the Sun is made up of almost all gases. As the Sun loses mass over time, it loses its internal gravity. The gases, under less gravitational pull, expand outward. This means that the Sun will be less dense as time goes on, but will actually get bigger.

## SWALLOWING UP PLANETS

If the Sun continues using its fuel supply of hydrogen at its present rate, the Sun will last another 5 billion years. During the last couple hundred million years of the Sun's expected life cycle, the Sun will expand, cool, and turn orange and then red. It will eventually grow so large that it will engulf the planets Mercury, Venus, and Earth. As the end of Earth approaches, the Sun will no longer be a yellow-white ball but a gigantic red orb that fills most of the sky.

## THE EXHALE

At the end of its life, the fight between gravity and pressure that raged during the Sun's entire 10-billion-year lifespan will be settled. It will not have enough gravity to hold itself together, and its outer shells will expand and spread a ring of material, called a planetary nebula, like a shockwave throughout the solar system. The outer planets of Jupiter, Saturn, Uranus, and Neptune may survive this phase in part, or they may be utterly destroyed.

## LONG LIVE THE SUN!

In time, the Sun will expand and disperse, but that doesn't mean that it's dead. Once the gas clears and the planetary nebula disperses, the Sun will emerge from this apparent cocoon and shine on as a white dwarf star, an object tremendously hot and bright despite being only about the size of Earth. This Sun-turned-dwarf-star will live on for billions or even trillions of years.

## REDUCE, REUSE, RECYCLE

Once a planetary nebula (the outer shell of a star) is created from the expansion of a star (like the Sun), it will fly through space and eventually mingle with the detritus of other exploded stars. When enough of these parts coalesce, they will create a new, star-forming nebula and may create many more stars. This is a stellar recycling process on an astronomical scale.

## YOU'RE A STAR. AND YOU'RE A STAR. AND YOU'RE A STAR . . .

The jewelry you wear, the fillings in your teeth, and even the iron in your blood can only have come from one place: a tremendous supernova. A long time ago a massive star exploded and created these elements. So if you go back in time far enough, each and every part of you was inside a really humongous star. You and everyone you know are truly super stars!

# CHAPTER 4

# HOT WORLDS

*Mercury and Venus*

A meteorologist on Mercury or Venus would have an easy job. Every day the weather is pretty much the same. Mercury doesn't have an atmosphere so it is sunny every day and free of storms. As the closest planet to the Sun, it is baked during the daytime and reaches 800°F. But once the Sun sets, the surface temperature plummets almost immediately to −280°F because there isn't an atmosphere to hold in the heat. Think about that; a temperature fluctuation of more than 1,000°F over the course of a Mercurian day.

The temperature on Venus is even more predictable. Every day is cloudy and 900°F. Daytime, nighttime, on the equator, or at the poles, it is always 900°F and cloudy. These clouds dominate the weather on Venus by trapping the Sun's energy and distributing it over the entire Venusian surface. Of course, a visitor *would* need to watch out for the occasional rain shower, but an ordinary umbrella wouldn't do the trick. On Venus, when the clouds open up, they rain sulfuric acid instead of water.

Let's take a closer look at the two planets closest to the Sun and the unmanned spacecrafts that paid them a visit.

# MERCURY

## GOOD THINGS COME IN SMALL PACKAGES

Mercury is the closest planet to the Sun. And at only 3,032 miles in diameter, Mercury is the smallest planet in the solar system (now that Pluto is no longer a planet).

## IF MERCURY IS SO INCLINED . . .

Mercury has the least circular, most eccentric, and most inclined orbit of the eight planets in the solar system. This means that, by percentage, Mercury varies its distance from the Sun more than any other planet. Mercury's distance to the Sun can range from 28.5 million miles at its closest to more than 43 million miles at its farthest.

## SHRINKING AND GROWING SUN

If you spent one year on Mercury, you'd notice the Sun significantly change size. When Mercury is farthest from the Sun (called aphelion), the glowing yellow disc would cover 1.1 degrees of the sky. But at its closest approach to the Sun (called perihelion), the blazing orb would seem to take up 1.7 degrees of the sky—an increase of more than 50 percent.

## A CLOSE-UP VIEW

If you were standing on Mercury at perihelion, the Sun would seem incredibly big and bright. The Sun would look more than nine times larger and shine nine times brighter than your Earthly view of the Sun during Earth's perihelion.

## SEVEN DEGREES OF INCLINATION

If you look at the solar system from afar and map out the planetary orbits in three dimensions, you'd see that all of the planets circle the Sun in almost the same plane. Imagine the planets are marbles circling the Sun on a flat table—that's almost how it is in the solar system. Five of the seven other planets orbit the Sun within 3 degrees of Earth's orbital plane (inclination). Venus is inclined a little over 3 degrees off from that of Earth, and Mercury's orbit is inclined by 7 degrees.

## LEAST TILTED PLANET

Even though Mercury has an inclined orbit, it is the least tilted planet. Unlike Earth, whose 23.5-degree axial tilt causes its seasons, Mercury is tilted only about $\frac{1}{30}$ of a degree to its orbital path around the Sun, and its seasons are caused by its varying distance from the Sun.

## COLD AS ICE

Even though some portions of Mercury's surface can reach temperatures of more than 800°F during the day, other parts still have ice. Deep down in craters at the north and south poles of Mercury, the Sun never shines. At the bottom of those craters lies water still in a frozen state.

## MAGNETIC MERCURY

Mercury is the second-densest planet behind Earth. It has a molten iron core that makes up a large percentage (42 percent) of its total volume. The fact that tiny Mercury has not completely cooled off was a surprise to astronomers. This molten core helps generate a weak magnetic field (something the larger planets, Mars and Venus, lack).

## LIGHT WEIGHT

You could stand on Mercury because it's made of solid rock. The gravity on Mercury is 38 percent what it is on Earth, which means that someone who weighs 200 pounds on Earth would weigh a mere 76 pounds on Mercury.

## ALMOST NO ATMOSPHERE

Mercury has an atmosphere, but it is almost nonexistent. What little atmosphere exists is in constant flux. Mercury is such a small planet that gases escape the pull of gravity and jet into space. However, particles from the Sun combined with dust kicked up from meteorite impacts sometimes generate an extremely tenuous atmosphere.

## THE DARK PLANET

Mercury is dark. Of all the planets in the solar system, its dusky, rocky surface reflects the least amount of visible light. Mercury is not quite as dark as coal, but it is similar to the darkness of blacktop or the pavement on a well-traveled highway.

## LOOK-ALIKES

Pictures of Mercury look an awful lot like images of the Moon. Mercury's gray, battered surface is the most cratered planet in the solar system. It sports scars from ancient impacts just like the Moon.

## FROZEN IN TIME

There is no wind or water erosion on Mercury and little to no geological activity to erase the impact craters. That means the surface of Mercury shows almost all of its violent, ancient history frozen in time. The Mercury you can observe today has stood pretty much unchanged for millions to billions of years.

## HITTING THE BULL'S-EYE

An object more than sixty miles across slammed into Mercury almost 4 billion years ago, creating Mercury's largest crater, Caloris basin. The impact also produced a ring of mountain ranges, called the Caloris Montes, that are almost 1,000 miles across and 2 miles high. From above, it looks like a bulls-eye made of radiating rings of rock.

## MERCURY METEORITES?

There may be pieces of Mercury on Earth. In 2013, a strange, green rock found in Morocco attracted the attention of geologists. It seems to be a meteorite, but no one knew where it originally came from. Some scientists believe that it bears the same chemical signatures that the MESSENGER spacecraft detected on Mercury, but the origin of this rock remains unknown.

## IT'S YOUR BIRTHDAY . . . AGAIN

Mercury takes only eighty-eight Earth days to orbit the Sun, which means if you lived there you'd have a birthday every eighty-eight days. Talk about a party planet!

## SLOWLY ROTATING MERCURY

For a long time astronomers either did not know how long it took for Mercury to rotate, or they assumed that one side always faced the Sun, which would mean that Mercury's day and year would be the same (eighty-eight Earth days). In the 1960s, astronomers finally figured out that it takes fifty-nine Earth days for Mercury to rotate once.

## SEEING YOUR HOME

If you visit Mercury, you'd be able to see Earth regularly with the naked eye. And it would be bright and blue. Earth would appear to be the second-brightest "star" at night behind the planet Venus.

## A COMPLICATED SUNRISE

The time from sunrise to sunset on Mercury is pretty complicated. If you were standing on Mercury, the period from sunrise to sunrise—what you normally think of as a full day—is twice as long as its year. The planet rotates so slowly that you would only see one sunrise for every two trips around the Sun (176 Earth days).

## WAIT, WHAT TIME IS IT?

Mercury's highly elliptical orbit and slow rotation rate plays havoc with timekeeping. In some places and some occasions on Mercury, you could see the Sun rise, go across the sky, stop, go backward, and set in the same place it rose.

## THE JUMPY MESSENGER

Assyrian astronomers in the fourteenth century B.C. were the first to name the planet Mercury. They called it UDU.IDIM.GUD.UD or "the jumping planet," most likely because it jumped so quickly from being visible in the morning sky to the evening sky and back again. Ancient Babylonians referred to Mercury as Nabu, their messenger god and god of writing. Part of Nabu's nature was adopted into the Greek god Hermes and the Roman god Mercury.

## MERCURY'S DAY

Wednesday is Mercury's day. Each day of the week was originally attributed to a deity, and "Wednesday" comes from the Norse god Odin (who shared a lot of similarities with Mercury). The word *Wednesday* emerged from the Old English word *wodnesdaeg*, or "Wodin's day." Mercury's association with Wednesday is a lot more evident in the Romance languages. For instance, Wednesday is called *Mercredi* in French and *Miercoles* in Spanish.

## FASTEST PLANET IN THE SOLAR SYSTEM

Mercury is speedy. It travels at an average rate of almost 106,000 miles per hour in its orbit around the Sun. However, it moves even faster when it's closest to the Sun, at a speed of more than 130,000 miles per hour. It is by far the fastest planet in the solar system.

## DON'T BE SHY

Mercury is the toughest planet to see clearly from Earth since it is always near the Sun in the sky. Since you can only see it in the sky just after sunset or just before sunrise, Mercury makes an elusive target for any ground-based telescope.

## PRETTY IN PINK

Despite its dark gray surface, from the surface of Earth Mercury often looks pink in color. This is because you can only see it when it is low in the sky, and its reflected light travels through a lot more of Earth's atmosphere to reach your eye. This is the same reason that sunsets look so red.

## DOUBLE TAKE

The ancients regularly saw Mercury, but some cultures considered it to be two separate objects: one that appeared in the morning sky and a wholly different one that came in the evening. Eventually they figured out that both objects were one and the same: Mercury. They must have asked themselves, "Wait, how come I never see the morning guy and the evening guy at the same time? Or even the same month?"

## SHRINKING WITH AGE

As it aged, Mercury shrank. Since its formation over 4 billion years ago, it has cooled and its exterior condensed by about 4.4 miles. This created a lot of geologic features on the planet that give it a crumpled texture.

## MERCURY'S TAIL

The solar wind scours Mercury's surface and rips atoms off the planet. This creates what astronomers call an exosphere. Some of these atoms fall back to the planet, but others flow into space and give Mercury a super-faint tail that points away from the direction of the Sun.

## THIN-SKINNED

The liquid iron core of Mercury makes up more than $2/5$ of its total volume (the highest proportion of any planet). Its core is more than 1,100 miles thick, while the outer layers of rock that comprise the mantle and crust are only 400 miles thick.

## MOONLESS

Mercury doesn't have any moons. It's such a small planet that it would have trouble attracting a passing asteroid into orbit around it. Plus the Sun's proximity and mass would disrupt a moon's motion and make it extremely difficult for anything to achieve a stable orbit around Mercury.

## FROM MARINER TO MESSENGER

Mercury has had only two visitors from Earth. From 1974–1975, Mariner 10 rocketed past the planet three times. And the MESSENGER spacecraft flew by Mercury twice in 2008 and once in 2009 before settling into Mercury's orbit between 2011 and 2015.

## BAD TIMING

Before 1974 astronomers had no idea what Mercury looked like up close. When Mariner 10 made its flybys of Mercury, it was able to image 45 percent of the planet. On its second and third passes, Mariner 10 took more pictures. But each time it flew by, pretty much the same half of the planet was in daylight while the other half remained in shadow, which meant that the other half of Mercury sat unseen for more than three more decades.

## GLOBE OF MERCURY, NOW AVAILABLE

A global map of Mercury was not completed until 2011 when NASA's MESSENGER spacecraft charted the entire surface of the planet. MESSENGER is actually an acronym for MErcury Surface Space ENvironment GEochemistry and Ranging, and it allowed astronomers to learn even more about the closest planet to the Sun.

## A TOUGH NEIGHBORHOOD

It's really tough to orbit Mercury. If you fly there directly, the Sun's gravity along with Mercury's elongated orbit makes staying there almost impossible. So missions like MESSENGER had to sneak up on Mercury. MESSENGER launched off Earth and flew around the Sun a few times to get the best approach to Mercury. In its six-year journey to Mercury, it shot past Earth once, Venus twice, and Mercury three times before settling into a stable orbit.

## LEAVING ITS MARK

The MESSENGER mission ended on April 30, 2015, when the spacecraft ran out of fuel and crashed on the surface of Mercury. Unfortunately, it came to rest on the nighttime side of Mercury, so astronomers could not observe the impact and its immediate aftereffects. However, based on its final speed and trajectory, MESSENGER likely made a crater fifty-two feet wide and can be found strewn about an area on Mercury called Suisei Planitia.

## BEPICOLOMBO: THE NEXT VISITOR

The next scheduled visitor to Mercury is a joint operation between the European Space Agency and the Japanese Aerospace Exploration Agency. The unmanned mission is called BepiColombo, and like MESSENGER it will take a similar roundabout path—after leaving Earth it will fly around the solar system for seven years before finally entering into an orbit around Mercury where it will study the planet and its magnetic field.

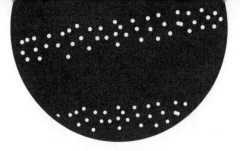

# VENUS

## LIGHT BRIGHT

From Earth, Venus is the brightest star-like object in the night sky. Often mistaken for a plane or UFO, Venus is so much brighter than any other star or planet that its appearance is startling to Earthlings.

## IT'S JUST A PHASE

When viewed through a telescope Venus exhibits a slowly changing array of phases. Venus appears like a full disc when it is farthest from Earth and on the other side of the Sun. But as it rounds the Sun and comes closer to Earth, sometimes you can see only half of Venus illuminated, like a half-moon. The closer Venus approaches Earth, the more it resembles a crescent. It may look like a little crescent Moon in a telescope, but it is actually a crescent Venus.

## BRILLIANT CRESCENT

From Earth, a crescent Venus is brighter than a full Venus. That's because when Venus is full, it is on the opposite side of the solar system—about 160 million miles away. But when it's in its crescent phase, Venus is only about 40 million miles from Earth and appears brighter.

## WON'T YOU BE MY NEIGHBOR

Although sometimes it is farther from Earth than Mercury or Mars, Venus can be the closest planet to Earth. At its closest to Earth, when Venus comes between Earth and the Sun, it is about 26–27 million miles away. Mars can only get as close as 34.6 million miles, and the closest Mercury can get to Earth is about 48 million miles.

## CASTING SHADOWS

Venus can be so bright in the night sky that it can cast shadows on Earth. Well, to clarify this a bit, on a crystal clear evening, when there is no moonlight, no ambient city lights, when Venus is at its brightest, and the Sun has long set, you might be able to detect Venus casting a faint shadow. Cameras can definitely pick it up, and someone with particularly good eyesight can isolate a subtle outline caused by the light of Venus casting a shadow.

## TWISTED SISTER

Venus is often called Earth's sister planet since it is almost the same size as Earth. Venus's diameter is 7,580 miles, while the diameter of Earth is 7,918 miles. But beyond size, the two planets have very little in common. If Venus is Earth's sister, it is one twisted sister.

## SO, SO DRY

Although Venus is smaller than Earth, it has much more land. While more than 70 percent of Earth is covered by water, Venus is 100 percent land. This gives Venus more than three times as much land area as Earth.

## UNDER PRESSURE

Venus is made of rock, so you could stand on its surface. But you wouldn't want to. The surface of Venus is always about 900°F, partially because of the ever-present clouds. They not only trap in heat, but they also are extremely heavy. On Earth, the atmosphere pushes down on you very comfortably. You've lived with it so long that you don't even notice it. But on Venus, the atmosphere would exert such a force on you that your body would crumple under the weight.

## THE 10 PERCENT

On Venus you'd weigh 90 percent of what you weigh on Earth. So a 200-pound person on Earth would only weigh 180 pounds on the surface of Venus. Out of the other three rocky planets (Mercury, Venus, and Mars), Venus's gravity will feel the closest to what you experience on Earth.

## THE DAYS ARE LONG

A day on Venus is longer than its year. While Earth completes one spin each day, Venus rotates so slowly that it takes 243 Earth days to spin one time. Meanwhile, Venus is closer to the Sun than Earth, so its year (the time it takes to orbit the Sun once) is only 225 Earth days. Venus is the only planet in the solar system with a situation like this.

## VENUS'S RAPID, CIRCULAR ORBIT

Venus is the second-fastest planet in the solar system. It travels at an average speed of 78,000 miles per hour, and because Venus has the most circular orbit of any planet, its velocity does not change much over the course of its trip around the Sun.

## THE UPSIDE-DOWN PLANET

Venus is tilted in a way that is completely different from any of the other seven planets. It appears to be upside down and spins backwards. If Earth spun in the same direction as Venus, the Sun would rise in the west and set in the east.

## A GLANCING BLOW

Scientists believe that a large object may have struck Venus at such a glancing angle that it flipped the planet and slowed its rotation. This collision theory has the most scientific evidence behind it at present, but it is hard to envision an entire planet getting knocked over.

## WHERE'S THE MOON?

Venus doesn't have any moons. But if a giant impact created the Moon and astronomers believe that Venus also suffered a monumental collision, where is Venus's moon? This is yet another Venus mystery. Perhaps the collision that created the Moon was much different in size, scale, and effect than the one that smacked Venus all topsy-turvy.

## VENUSIAN METEORITES

Currently there are no known meteorites on Earth that came from the planet Venus. Out of all the inner planets, rocks blasted from the surface of Venus would have the most difficult time journeying to Earth due to Venus's strong gravity. However, astronomers know that Venus was bombarded by large meteorites, and in theory, rocks from Venus could be on Earth right now.

## LONG DAY, GETTING LONGER

Venus is winding down. Based on data from the Magellan and Venus Express orbiters, the planet isn't spinning as fast as it once was. Its daily rotation rate is slowing. And the change is dramatic (astronomically speaking). In only twenty years, Venus's day has gotten longer by 6.5 minutes.

## EIGHT YEARS FROM NOW

The next time you see Venus in the sky, remember where you were and what it looked like. If eight years later you return to the same location on the same day, Venus will be in almost exactly the same place in the sky in front of the same constellation.

## EARTH: 8-VENUS: 13

Earth and Venus have an 8:13 resonance. This means that when Earth makes eight orbits of the Sun (eight years), Venus has traveled almost exactly thirteen times around the Sun. That's why Venus appears in the same place in the sky every eight years.

## ANCIENT VENUS WATCHERS

Ancient Babylonian astronomers documented Venus's risings and settings and compiled the data on a clay tablet called the Venus tablet of Ammisaduqa. Dated to the seventh century B.C., it is one of the oldest surviving astronomical records.

## AFFINITY FOR VENUS

The Maya who lived in the Yucatan peninsula had a special fascination with the planet Venus. In the Dresden Codex, a Mayan book collected by the Spanish explorers in 1519, Mayan astronomers laid out a Venus table— predictions for the positions of Venus during its five cycles of 584 days.

## FEMALE DIVINITY

According to Roman mythology, Venus was the goddess of beauty. The symbol for this planet is a circle atop a cross, the same symbol used for the female sex.

## VENUSDAY!

In Europe, Friday was associated with the planet Venus. In Spanish, Friday is called *Viernes*. In French, it is *Vendredi*. The English name *Friday* may come from one of two Norse goddesses, Frigga or Freya, who were each linked to the planet Venus at one time or another in Norse mythology.

## VENUSIAN STARGAZERS

If it weren't cloudy on Venus's surface, the Sun would appear about 1.77 times larger than it does from Earth. Earth would be the brightest star-like object in the sky, and if you have good eyesight, you might be able to see the Moon as a little speck of light next to it.

## NEVER A CLEAR NIGHT

It's always cloudy on Venus. This planet would be the worst place to be an astronomer. The ever-present clouds would block your view of the stars during the long Venusian night and even obscure the Sun during the blazing hot days. If you like stargazing, Venus is not the place for you.

## STINKS TO BE FIRST

Many of the first unmanned missions to Venus were met with a string of failures. Twelve of the first thirteen missions (from the United States and Soviet Union) failed during launch or en route to Venus. The Soviet craft Venera 3 lost control and crashed into Venus, becoming the first human-made object, however broken, to reach another planet.

## FLYBY MISSION

The lone mission that did not fail was NASA's spacecraft Mariner 2. In December 1962, it passed within 22,000 miles of the surface of Venus, making it the first spacecraft to fly by any planet. Mariner 2 confirmed Venus's sweltering temperature and crushing pressure but could not peer through the clouds that reached as high as thirty-seven miles.

## A HARSH LANDING

The first spacecraft to successfully land on Venus (or any other planet) and send information back to Earth was the Soviet craft Venera 7. However, the landing did not go according to plan. During the descent, Venera 7's parachute failed to open and the spacecraft smacked into the Venusian surface and probably bounced onto its side. Russian scientists were amazed to find that the craft still functioned for twenty minutes even in its crumpled state and informed scientists about the hellish surface temperatures and pressures found on Venus.

## VENERAS TO VENUS

Between 1972 and 1984, the Soviet Union flew a fleet of mostly successful Venera missions to Venus. Veneras 8 through 14 achieved some, if not all, of their mission goals, including landing safely (not an easy task) and transmitting information and pictures back to Earth. It took the Venera missions about four months to fly from Earth to Venus, but their instruments each collected, on average, less than one hour of data.

## TWO HOURS ON VENUS

Venera 13 was the Soviet mission that survived the longest through the heat and pressure of the Venusian surface. It sent back data for 127 minutes, which wasn't too shabby since it was only designed to last for thirty-two minutes.

## GETTING IN ORBIT

In 1990, NASA's unmanned Magellan spacecraft became the first mission to orbit the planet Venus. It hitched a ride into Earth's orbit in the payload bay of the space shuttle *Atlantis* in 1989. Astronauts released it into space where it rocketed away from Earth, took a roundabout fifteen-month journey to Venus, and settled into orbit.

## THROUGH THE CLOUDS

From 1990–1994, the unmanned Magellan spacecraft used radar to peer beneath the thick clouds and map the surface of Venus. Over the course of thousands of orbits, Magellan imaged 98 percent of the surface and delivered the highest-resolution maps of Earth's sister planet. Before then, Venus's surface was largely shrouded in mystery.

## FAREWELL TO MAGELLAN

The Magellan mission ended on October 13, 1994, when it was purposely crashed into Venus. As its fuel supplies dwindled, NASA engineers executed one final mission for the plucky spacecraft: study the clouds of Venus from within. During its plunge, Magellan continued to beam back information to Earth until it was ripped apart by the friction with the atmosphere. No one knows where its pieces came to rest or if they even remained sufficiently intact to hit the ground.

## WHERE ARE ALL THE CRATERS?

Venus's surface has surprisingly few impact craters. Unlike Mercury and the Moon, Venus shows a lot of geological activity including many more volcanoes than those found on Earth that wiped away many of the scars from meteor impacts. However, there are still some craters on Venus, and one of them, Addams crater, is almost sixty miles wide with a distinct rim around it. But the impact melted so much of the rock that it flowed out one side of the crater in a 373-mile-long lava river. The lava has since cooled leaving a bright trail of new rock atop the darker, older plain.

## VENUSIAN VOLCANOES

Despite having such a thick atmosphere, Venus shows little to no signs of wind erosion and no signs of water. However, large areas of Venus show signs of ancient volcanic activity. There are hardened lava plains, lava fields, and lava channels as well as several types of dormant volcanoes.

## ATOP MAXWELL MONTES

The highest mountain range on Venus is called Maxwell Montes, which towers 4 miles above the nearby plain called Lakshmi Planum and soars 6.8 miles above the average elevation of the planet. Atop Skadi Mons, the peak of the Maxwell Montes, are very bright patches that may be metallic snow. That snow is definitely not water ice, but is most likely a combination of lead and bismuth sulfide that condensed out of the atmosphere.

## THE TICK

One Venusian volcano looks like a tick; it has a rounded top and lines radiating out of it like the legs of an insect. Geologists call these features scalloped margin domes. However, one image taken by the Magellan spacecraft definitely looks more like a blood-sucking parasite than a large volcano.

## SURF'S UP!

The latest manmade visitor to Venus was the ESA mission Venus Express. After a five-month trek from Earth, Venus Express entered orbit around Venus in April 2006. It studied the Venusian atmosphere and climate for eight years. Toward the end of its lifespan, Venus Express surfed in and out of the upper atmosphere. Eventually it ran out of fuel, and one time after going down below the clouds it never came back up.

## IS THERE LIFE ON EARTH?

The spacecraft Venus Express studied Earth from afar. The goal was to see if it could detect signs of life on Earth from millions of miles away. The experiments Venus Express used are being fine-tuned by astronomers on current searches for life on planets outside the solar system.

# MERCURY AND VENUS: FLYING CLOSE TO THE SUN

### A MERE TWINKLE

Mercury and Venus (and all of the planets, for that matter) shine by reflecting the light of the Sun. They do not generate any significant light of their own. Sunlight travels millions of miles, bounces off their surfaces, and journeys millions more miles to reach your eyes. From Earth, they look like bright stars that barely twinkle.

### THE GREATEST ELONGATION

It's easiest to see Mercury and Venus in the nighttime sky when they create the most angular separation from the Sun. This is called greatest elongation. When either planet is at greatest eastern elongation, you can see it just after sunset. When they're at greatest western elongation, you can see Mercury and Venus before sunrise.

### TRANSIT SYSTEMS

Mercury and Venus are the only two planets that can come between Earth and the Sun. Since Mercury and Venus are too small and too far away to block the entire Sun out from Earth's perspective, they look like little black dots slowly going in front of the blazing Sun. Astronomers call these mini eclipses transits.

## 'TIS THE SEASON

Each planet has two transit seasons. Transits of Mercury can only occur in the months of May and November, and transits of Venus happen only in June and December. These transits are much rarer than solar or lunar eclipses. From your perspective on Earth, Mercury crosses in front of the Sun thirteen or fourteen times per century. The next ones will be on November 11, 2019, and November 13, 2032.

## DON'T HOLD YOUR BREATH

Transits of Venus are really rare and follow a complex pattern. Every 243 years, you'll see four transits of Venus separated like so: transit number one occurs, then transit number two comes 8 years later, transit number three is 105.5 years after that, transit number four happens 8 years later, and finally 121.5 years after that the cycle repeats itself. There were two transits already in the twenty-first century (June 8, 2004, and June 5, 2012). That means you'll have to wait a while to see the next one . . . on December 10, 2117.

## MERCURY FROM MARS

Sun-watching satellites have captured images of Mercury and Venus crossing in front of the Sun from their lofty perches in space. The Curiosity rover on Mars recorded a transit of Mercury from the Red Planet in 2014. From that distance, Mercury looked like an even smaller dot moving across the Sun.

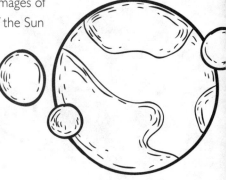

# CHAPTER 5

# IF YOU WERE A MARTIAN

A Travel Guide to the Red Planet

Mars is mysterious. And it's a little bit of a tease.

If you were an ancient astronomer, the planet Mars's ruddy appearance in the skies would have given you pause. Its odd motions through the constellations would have vexed you. If you were a Renaissance astronomer, the invention of the telescope merely added to Mars's mystique. At times, when you squinted through the eyepiece, you got glimpses of the Martian surface so clear that you'd think, "I see it." But then the features quickly changed. If you were a nineteenth-century astronomer, you built a bigger telescope and looked at Mars when it was really close to Earth, seeing lines on Mars that you called channels. This eventually led to other astronomers figuring that if the channels were actually "canals," an intelligent race of beings must have designed and built them. And if an intelligent race of beings was on Mars, then you had to do whatever you could to find them. So modern astronomers flew spacecraft by Mars, around Mars, and landed on Mars. They used rovers to search and dig for life. They found water, which indicated that life might not be there now but maybe it was, so they looked for fossils. And then they looked underground . . .

It seems like finding life on Mars is always one step away. Will it take a giant leap, or space flight, and require humans to travel to Mars to uncover its mysteries? Or will Mars guard its secrets, teasing those who want to unravel its mysteries for another 100 years? While there's a lot to learn about the Red Planet, a lot is already known as well. In this chapter you'll discover what modern astronomers know and how they are going to learn more.

# MARS BASICS

### NUMBER FOUR

Mars is the fourth planet from the Sun and also the fourth-fastest planet. It travels at an average rate of about 53,600 miles per hour around the Sun, but its speed varies based on its changing distance from the Sun. When Mars is closest to the Sun it speeds up to over 59,000 miles per hour, and when it swings around to its farthest point from the Sun it slows down to about 49,000 miles per hour.

### AN EARTH-LIKE TILT

Mars is tilted on its axis by 25 degrees. Of all the planets in the solar system this is the most similar tilt to Earth's 23.5 degrees. This tilt gives Mars four seasons just like Earth. But a year on Mars is 687 Earth days, so each Martian season lasts almost twice as long as a season on Earth.

### EXTREME SEASONS

Mars's eccentric orbit throws an extra wrinkle into its seasons. Unlike Earth, Mars's distance from the Sun changes dramatically. When it's closest to the Sun (perihelion) it is 128 million miles away. When it's farthest from the Sun (aphelion) it is almost 155 million miles away. When a winter season corresponds with aphelion, you get an extreme winter. And a summer at perihelion can make a day on Mars warmer than some places on Earth.

### ANOTHER DARK PLANET

Mars is a much darker planet than Earth. Its rocks, on average, reflect less than half the amount of sunlight as the surface of Earth. As planets go, only Mercury is darker than Mars.

## BRIGHT BEACONS OF LIGHT

The ice caps of Mars are very bright compared to the rusty rocks of its temperate and equatorial zones. In fact, the north and south polar ice caps are so bright that when Mars is closer to Earth you can see one or both of the ice caps stand out like white beacons through a backyard telescope.

## THE REDDISH PLANET

Mars is known as the Red Planet since, from Earth, it appears redder than most stars in the sky. But its color also changes to golden, orange, brown, or pink based on the sky conditions on Earth and the weather on Mars. Dust or haze in Earth's atmosphere can wash out the red color, as can giant dust storms covering huge swaths of the Red Planet.

## THE TWENTY-SIX-MONTH SHINE

Mars appears biggest and brightest when it is closest to Earth. This occurs when Mars is on one side of the sky, the Sun is on the opposite side, and Earth is in between. Astronomers call this opposition, and it happens about every twenty-six months.

## AS CLOSE AS IT GETS

When Mars is at perihelion (its closest point to the Sun) and opposition to Earth, it shines extremely bright in the nighttime sky. It outshines all the stars and planets except Venus. This configuration happened in 2003 when Mars was closer to Earth than it had been in 60,000 years.

## NO SUCH THING AS AN EQUAL OPPORTUNITY OPPOSITION

The extremely close opposition of 2003 brought Mars to within 34.6 million miles from Earth. However, during its opposition in 2012, Mars only got as close as 63 million miles.

## BLURRY AND BORING

Mars is the most popular planet and the most requested planet to see in a telescope. Unfortunately, through a telescope Mars is not as exciting to behold as Jupiter or Saturn. It usually just looks like an orangeish, pinkish disc. But during close approaches you may be able to see a white circle marking either the north or south polar cap, and dark markings on its surface. These dark rocks are blurry and seem to change over time when you squint long enough through a telescope eyepiece.

## MARS-MOON HOAX

Starting in 2004, an e-mail hoax circled inboxes around the globe pronouncing that Mars would look as big as the full moon. Although that would be amazing, it is impossible. Even when Mars was closest to Earth in 2003, its diameter appeared to be seventy-six times smaller than that of the Moon.

## RED PLANET RETROGRADE

When the ancients charted Mars's motion across the background stars, they noticed something strange. It went steadily forward for hundreds of days, then slowed down and went backward relative to the constellations. Astronomers call this backward movement retrograde motion. When astronomers map Mars's apparent motion in the sky year after year, its path looks like a loop-the-loop or a zigzag.

## CIRCLE UP!

Mars's motion really confused the ancient astronomers. In A.D. 150, Greek astronomer Claudius Ptolemy made a detailed map of the universe in a book called the *Almagest*. In it, he placed Earth at the center with all of the planets circling around it. To account for Mars's peculiar loop-the-loops, he added circles within circles that seemed to predict Mars's position with great accuracy—even though Ptolemy thought the Red Planet went around Earth and not the Sun.

## THE SUN STANDS ALONE

In 1504, Polish astronomer Nicolaus Copernicus did not find Mars where Ptolemy said it should be. He took a look and discovered that everything was in its proper place if you put the Sun in the center of the solar system instead of Earth. This heliocentric theory revolutionized astronomy, and the Red Planet provided the clue.

## GET ON THE ELLIPTICAL

German astronomer Johannes Kepler used Mars observations from Danish nobleman and astronomer Tycho Brahe to figure out that the Red Planet did not orbit the Sun in a perfect circle. None of the planets do. Instead, planetary orbits are elliptical in shape, and Kepler solved this puzzle in the early 1600s.

## GIVE ME LAND, LOTS OF DRY LAND

Mars is 4,212 miles in diameter—just a little over half the width of Earth. However, both planets have about the same amount of dry land. While only 29 percent of Earth's surface is not an ocean, sea, lake, or river, all of Mars is solid, dry land.

## MARS AND MERCURY

The gravity on the surface of Mars is about 38 percent that of Earth's, as is Mercury's. Mercury is smaller than Mars, but it is also denser. Additionally, since Mercury is smaller, you'd be closer to its center of mass than you would be on the surface of Mars.

## A DAY ON MARS

A day on Mars is only slightly longer than a day on Earth. It takes Mars twenty-four hours, thirty-seven minutes, and twenty-two seconds to spin once compared to Earth's twenty-four hours.

## WHEN'S BEDTIME AGAIN?

Initially, NASA scientists on Earth who drove the Mars rovers Spirit and Opportunity adjusted their wake and sleep schedules to correspond to the Martian day. They went to work and drove the rovers when it was daytime on Mars, but not necessarily daytime on Earth. So every day they went in to work thirty-seven minutes later than the previous day. Eventually the schedule became so disorienting to their body clocks that they abandoned the Martian schedule. NASA rotated drivers more regularly so they could keep close tabs on an Earthly day.

## LIFE ON MARTIAN TIME

The first astronauts to visit Mars will have no choice but to adapt to the planet's twenty-four-hour, thirty-seven-minute, and twenty-two-second daily schedule. They'll have to fine-tune their body clocks to the rising and setting of the Sun over the Martian landscape.

## OLYMPUS MONS IS SO BIG . . .

The largest volcano and mountain on any of the planets in the solar system is Mars's Olympus Mons. How big is it? Olympus Mons is so big that it's about three times taller than Mount Everest, and if it were placed in the southwestern United States, it would cover the entire state of Arizona.

## EXTINCT OR JUST RESTING?

Olympus Mons is a shield volcano created by layers upon layers of lava expelled from its caldera over millions of years. Underneath Olympus Mons, in an area called Tharsis, is a hot spot that raised several other large volcanoes. These volcanoes look extinct, but new data suggest otherwise. Some geologists think that Tharsis may be merely dormant while the hot spot gathers energy. It could become active again!

## THE GRANDEST CANYON

Volcanic activity on Mars created one of the largest canyons in the solar system. Valles Marineris, at over 2,500 miles long, 120 miles wide, and 7 miles deep, looks like a deep gash across the surface of the Red Planet.

## ONE HUMONGOUS CRATER

Mars's Borealis basin may be the largest crater in the solar system. Covering 40 percent of the planet, this low, flat bowl around Mars's north pole may have been caused by a large meteor slamming into the planet long ago.

## BRRR . . .

Mars is generally a cold planet, but for several days during the winter of 2015 the high temperature on Mars was warmer than the high temperature in Midwestern cities like Cincinnati, Ohio, and Chicago, Illinois. While high temperatures that winter hovered around 0°F, Mars got into the single digits. The highest air temperature ever recorded on Mars was 95°F.

## MARS AFTER DARK

The average temperature on Mars is −60°F. Once the Sun sets, the heat escapes the thin atmosphere almost instantly. The lowest recorded temperature on the Martian surface was −166°F.

## NONMAGNETIC MARS

Unlike Earth and Mercury, Mars doesn't have a magnetic field. This means that charged solar particles can bombard the surface unabated. However, scientists found evidence that Mars had a magnetic field in its distant past—about 4 billion years ago. Where did it go? That is still being debated.

## TIW FOR TUESDAY

You might be surprised to learn that Mars influenced the day of the week you know as Tuesday. The ancient Norse god of combat was named Tiw. Tiw drew a striking comparison to the Roman deity Mars and thus was likened to the planet of the same name. So Tuesday is really Tiw's day. The connection between Mars and Tuesday is more evident in other languages. In Spanish, Tuesday is *Martes*, and in French it is *Mardi*.

## MEN ARE FROM MARS

The symbol for the planet Mars is a circle with an arrow sticking out of it at a diagonal. This is also doubles as the symbol for the male sex.

# MARS MISSIONS

Missions to Mars are dicey. Out of the more than forty unmanned missions sent to the Red Planet by various countries, only about half of them were successful. Some crafts flew out of control, some crashed, and others were lost in space.

## THE RED MENACE

The unmanned space probes the Soviet Union sent to Mars had a very high failure rate. Their first ten missions to Mars failed during launch or en route. Their eleventh try, a mission inappropriately titled Mars 2, made it to Mars and circled the planet from 1971 to 1972. Mars 2 included a lander with a small rover inside that detached from the main orbiter and headed down to the surface. Unfortunately, during descent, there was a system failure and the lander crashed into the Martian plains. Still, the Mars 2 lander was the first Earthly object to reach Mars.

## CRASH TEST DUMMIES

American missions to Mars had their share of problems. Mariner 3, NASA's first attempt to reach Mars in 1964, never made it. And in 1998 and 1999, NASA had back-to-back failed Mars missions. The Mars Climate Orbiter got too close to Mars and broke into pieces above the planet. Then the Mars Polar Lander crashed into the red soil of Mars when its descent engines failed to fire.

## LET'S TAKE A LOOK

On July 14, 1965, the American craft Mariner 4 flew about 6,000 miles above the surface of Mars and sent twenty-two pictures back to anxious scientists. Astronomers finally had their first close-up images of the Red Planet. Mariner 4 revealed Mars to be an old, cratered, geologically dead world free from rivers, lakes, or oceans of water.

## FALSE POSITIVE

The first successful landers to the Red Planet were NASA's Viking 1 and Viking 2 spacecraft. In 1976, the Vikings relayed the first close-up images of the Martian surface, sampled the rusty soil, and looked for life. One of the Viking's soil samples tested positive! Unfortunately, it was a false positive. An unexpected chemical reaction fooled the sensors on the lander into thinking it had found life on Mars.

## WATCH IT!

Beyond Earth, Mars is the most watched planet. Currently the Red Planet has five orbiting spacecraft (2001 Mars Odyssey, Mars Express, Mars Reconnaissance Orbiter, Mars Atmosphere and Volatile Evolution (MAVEN), and the Mars Orbiter Mission) and two rovers that explore the planet's surface (Opportunity and Curiosity).

## MARS IN HIGH-RES

Every square mile of Mars has been mapped by various orbiting space-craft. The resolution on some of the pictures is so good that you can see boulders just a few feet across. And in some pictures, you can see the rovers and their tracks—from space!

## WATCH THE LIVE STREAM

Back in 1997, when the Internet was young, a Mars lander named Mars Pathfinder streamed images from its landing spot on the Red Planet. You could tune in daily to watch alien pictures come to your computer. Mars Pathfinder then released a little rover called Sojourner that rolled 330 feet over the Martian soil during its eighty-three days of life.

## SOJOURNER TO THE RESCUE

Sojourner, a small rover on Mars, became an instrumental plot device in two science-fiction movies. In both **Red Planet** and **The Martian**, stranded astronauts used Sojourner for spare parts in order to save their lives. In reality, Sojourner is still sitting where it stopped rolling, unless the dust storms moved it or buried it.

## THE WILD ROVERS

Two NASA rovers, named Spirit and Opportunity, landed on Mars in January 2004. During their descent, each spacecraft had to slow down rapidly so as not to crash into the Martian dust. They each employed a parachute, fired retrorockets, inflated air bags, detached the parachute, and then bounced to a stop on Mars.

## DRAGGING ONE WHEEL

One of the six wheels on the rover named Spirit stopped turning in March 2006. It limped along from place to place dragging its one wheel behind, and NASA scientists took advantage of the situation by studying minerals that Spirit uncovered as it plowed across Mars.

## RIP SPIRIT

Spirit rolled 4.8 miles on Mars from 2004 to 2009 until it became stuck in deep, soft soil. NASA engineers tried for nine months to maneuver the rover out of this sand trap but ultimately gave up on January 26, 2010. Spirit can still be found at a location called Troy, 14.6 degrees south latitude and 175 degrees east longitude.

## HOLE IN ONE

When the rover Opportunity bounced onto Mars's surface, it rolled into Eagle crater, a seventy-two-foot-wide hole in the Martian plain. When it came to rest in the dead center of the crater, from above it looked like a golf ball at the bottom of a cup.

## MARTIAN MARATHON

Once on Mars, rover Opportunity rolled out of Eagle crater, explored nearby Endurance crater, then made the long trek (nearly six miles over the course of twenty-one months) to Victoria crater. It climbed the Columbia Hills and is still rolling on Mars. Slow and steady, Opportunity took eleven years and two months to drive 26.2 miles, the length of a marathon. It has far exceeded its ninety-day warranty.

## A CAR ON MARS

The largest Mars rover, Curiosity, landed on the Red Planet in 2012. It is the size of a small car at ten feet wide, nine feet long, seven feet tall, and weighs in at just less than one ton (1,982 pounds).

## SEVEN MINUTES OF TERROR

When the Curiosity rover landed on Mars, it was too heavy to employ airbags for a safe landing. Instead, NASA engineers designed an elaborate system of parachutes, retrorockets, and a sky crane to gently set it down on the surface. During the descent, which NASA playfully called, "seven minutes of terror," the engineers held their breath. Everything worked flawlessly and Curiosity beamed back a message saying that it was A-OK.

## DIG DEEP

The Mars rovers leave tracks on the red soil wherever they go. In some places the tire tracks are so deep that the rovers stop, turn around, and investigate the small trenches that they gouged into the surface. Sometimes what lies below the surface of Mars is more interesting than what is above.

# HUMANS ON MARS

## SEVEN MONTHS OF SPACE TRAVEL

Sending a human to Mars is much more complicated than sending someone to the Moon.
It takes a spacecraft about seven months to fly
from Earth to Mars. That would mean you would
be weightless, exposed to additional solar radiation, forced to get along
with your crewmates in a confined space, and have to use the space toilet for seven months.

## EFFECTIVE BUT TOTALLY GROSS

Radiation exposure is the biggest danger for astronauts flying to Mars.
They will be bombarded by extra solar radiation for seven months during their journey to Mars and seven more months coming back. Spacecraft with appropriate lead shielding would be too heavy to launch.
Some scientists propose that astronauts could use water, food, and their
own bodily wastes as shielding. Why not line the inside of the spacecraft
with material that you will generate yourself?

## A TWENTY-SIX-MONTH TURNAROUND

An astronaut who lands on Mars wouldn't be able to come right back
home. During the seven months that it took to get from Earth to Mars,
Earth moved around the Sun and would be millions of miles further
away. You'd have to sit tight on Mars for about one year until Earth came
around the Sun and was close enough for you to start the seven-month
journey home. In total you could be gone twenty-six months. Faster
return trips may be possible in the future, but will require more energy
and utilize technology not yet thoroughly tested.

## ONE-WAY TRIP

If reaching Mars with a manned mission is the goal, why bother coming back? Why not send a crew on a one-way trip to Mars? They can create a settlement there and live out the rest of their lives on the Red Planet. There have been thousands of volunteers for just such a privately funded mission, called Mars One.

## COUPLE UP!

NASA scientists are currently debating what team of people would be best suited for a Mars mission. To help alleviate the social isolation that an extended space mission would entail, some psychologists suggest that it should be comprised of married couples.

## MUSCULAR MAYHEM

After spending seven months in the weightlessness of space, your muscles and bone mass would deteriorate without hours of daily exercise. Even with a rigorous onboard exercise regimen, you may find it difficult to adjust when you land, despite the diminished gravity of Mars.

## SUN SHINING BRIGHT

From Mars, the Sun would look about two times smaller than it does from Earth, but it is still bright enough to light up the thin Martian atmosphere, giving it a rosy glow. At noon, the sky is so bright that you could not see any of the other stars.

## SUNSET IN BLUE

A sunset on Mars is blue. During the day, the Martian sky changes from pinkish in color to brown to pink again, but when the Sun rises and sets you'd see a distinctly blue tint to the thin Martian atmosphere.

## LIKE BREATHING EXHAUST

Although Mars has an atmosphere, you can't breathe there. The air on Mars is about 100 times thinner than the air on Earth, and 95 percent of that is made up of carbon dioxide.

## GLOBAL DUST STORMS

Even though it has a thin atmosphere, Mars generates the largest dust storms in the solar system. Winds fueled by the extreme seasons can rage over huge swaths of the planet for months. Some dust storms can cover an entire hemisphere.

## DUST DEVILS

The Mars rovers have witnessed a variety of small wind events on the Martian surface since dust devils pop up frequently on the plains. Looking like mini tornados, dust devils can swirl unabated across the dry, desert landscape of Mars. Luckily the rovers have emerged from these storms unscathed.

## RED ROCKS

Mars looks red because its soil and rocks have rusted. The Martian surface is rich in iron that has oxidized over the years and turned a rusty color. However, other soil samples look green, brown, and golden.

## TO CAP IT OFF

Mars has two ice caps: one around its north pole and another that covers its south pole. About 70 percent of this ice is frozen water, which deposits there during Mars's long, dark winter. The ice caps can be hundreds of miles across, but they grow and shrink noticeably based on the planet's seasons.

## A DUSTING OF DRY ICE

It rarely snows on Mars, but when it does, the flakes are made of carbon dioxide: dry ice. This makes a lot of sense since the atmosphere, which generates the snow, is mostly carbon dioxide.

## WINDY FJORDS

During the spring and summer when an ice cap is exposed to sunlight, new weather patterns emerge. Winds stream down the mountains of ice at 250 miles per hour, carve deep grooves in the ice caps, blow dust to the temperate zones, and create high, thin cirrus clouds.

## UNMELTABLE ICE

Mars's ice does not melt into water and flow into rivers and streams because the atmosphere is so thin that liquid water cannot exist on the surface for very long. Unlike on Earth where there is a water cycle, Mars has a sublimation/deposition cycle. Ice on Mars sublimates—turns from a solid back into a gas—directly into the atmosphere.

## MARTIAN EROSION

Mars has many impact craters on its surface. Like Mercury, Earth, and the Moon, meteoroids and asteroids bombarded it since it formed more than 4 billion years ago. Mars does not show as many craters as the Moon, however, since the forces of wind and water have eroded many of the older craters away.

## TOTALLY TUBULAR

Mars has lava tubes. Millions to billions of years ago lava flowed quickly through tunnels in the ground and spread across the surface of large, flat volcanoes. After the flows stopped, the tubes drained and what remained are hollow, almost perfectly circular holes in the ground.

## STARGAZERS' PARADISE

Stargazing on Mars would be incredible. With very little atmosphere to look through, you'd see thousands of stars with the naked eye. And although Mars is far from Earth, it is not far enough to significantly change your perspective on the distant stars. That means you'd see the same familiar constellations from Mars.

## STARRY SUBSTITUTES

Since Mars is tilted differently than Earth, the polar north star would be different if you were stargazing on the Martian surface. If you landed in the northern hemisphere of Mars, Deneb, the brightest star in the constellation Cygnus, would be the closest thing to a polar star. It would still be about 10 degrees from the north celestial pole of Mars, but as Mars rotates, all of the stars in the sky would seem to circle a spot near Deneb. The polar south star on Mars would be the star Earthlings call Kappa Velorum, which resides only about 3 degrees from the south celestial pole on Mars.

## MARS'S SPEEDY LITTLE MOONS

Mars has two moons: Phobos and Deimos. The names of these moons bolster Mars's brand as a war-like deity. The word *phobos* means "fear," and *deimos* means "terror." With mean diameters of 13.5 miles and 7.5 miles respectively, they are about the size of small cities, but they are so close to the planet that they would appear as the brightest objects in the nighttime sky. And they appear to move rapidly. On a nice Martian evening you could watch Phobos and Deimos wander through the constellations.

## COLONIZING PHOBOS

Future exploration of Mars may include a stopover on Mars's moon Phobos. In fact, some planned, manned missions to the Red Planet include establishing a base on Phobos. From Phobos, astronauts can avail themselves of the weak gravity and shuttle to and from the Martian surface with relative ease. Furthermore, when they want to return home, blasting off of Phobos requires much less energy than breaking free of the much stronger pull of Mars's gravity.

## ON TO AN ASTEROID

Astronomers and astronauts want to go to Mars, but dangers abound. A smaller step for mankind would be to visit an asteroid first. Although most asteroids lie farther away from Earth than Mars, some, called near-Earth objects, would make a nice destination. Future astronauts could practice living in deep space, landing on an object, and getting back home in a few months. It could be a good test for both equipment and the humans.

## LOOKING BACK HOME

From Mars, Earth would be brighter than any of the stars, and occasionally you might be able to spy the Moon right next to Earth with the naked eye. Earth would be best seen before sunrise and after sunset (just like Earthly views of Mercury and Venus).

## TRANSIT OF EARTH

Peering into a Martian sky, you could see a rare treat when Earth goes right in front of the Sun, looking like a little black spot on the Sun. This rare treat is called a transit of Earth, and you still have time to outfit your rocket to Mars in order to see the next one on November 10, 2084.

# MARTIANS

### THE STRAIGHT AND NARROW

In 1877, when Mars was extremely close to Earth, Italian astronomer Giovanni Schiaparelli saw straight lines crisscrossing the ruddy surface of the planet through his telescope. He made a Mars map of what he saw and called the lines "canali," or channels.

### THE ULTIMATE MARTIAN HUNTER

In the 1890s, wealthy American businessman Percival Lowell used his fortune to build the Lowell Observatory in Flagstaff, Arizona, in the hopes of proving that Mars was inhabited. Lowell confirmed that there were lines on Mars and decided that they were "canals" (not "channels" as earlier described) built by intelligent creatures. Lowell popularized the idea of Martian life, and his assertions have been the backbone of science-fiction stories ever since.

### OPTICAL ILLUSIONS

There aren't any canals on Mars. The lines observed by astronomers Schiaparelli and Lowell were tricks of the eye—illusions caused by squinting through a telescope under high magnification at a distant, blurry object. Their minds filled in dark patches and connected them unknowingly. The canals were merely a compelling optical illusion.

## SAVING FACE

While scouting out potential landing sites for the Viking 2 lander, the Viking 1 orbiter surveyed a region of Mars named Cydonia, a flat plain with strange rock formations. Several looked like massive pyramids and one resembled a human face. The "face" was huge—about 1.2 miles across—which led people to believe that the face was a work of Martian intelligence. In the 1990s, when the Mars Global Surveyor flew over Cydonia it saw the face in a different light. It was just an undulating mountainous outcrop that only looked like a face through the low-res camera of Viking 1.

## FACE THE FACTS

There are many rock formations that resemble faces on Mars. One mountain range, called Libya Montes, looks like a man with a flattop haircut wearing a turtleneck sweater. And Galle crater on Mars bears a striking resemblance to a smiley face. You are hardwired to see patterns in seemingly random shapes. The phenomenon is called pareidolia, and it causes you to see pictures in clouds, facial features in granite facades, and symbols in works of art.

## BIGFOOT?

On the Martian surface, the rovers have had close encounters with rocks that look like things. One looks like a rat, another like a human bone, and one looks like Bigfoot walking through a desert. Although these Mars rocks look like life forms on Earth, they are just rocks.

## FOSSILIZED ALIENS?

In 1996, a Martian meteorite named ALH84001 crashed into Earth. NASA scientists made a strong case that objects embedded in this rock were fossilized life forms from another world. President Bill Clinton even issued a statement to the press on the potential discovery. However, most scientists now believe that this announcement was premature. The "fossils" in ALH84001 are more likely the result of geologic processes and not life forms.

## MARTIAN METEORITES

The Mars rovers have examined several meteorites on the surface of the Red Planet. The first discovery is called Heat Shield Rock, since the Opportunity rover found it near the rover's discarded heat shield. In 2014, the Curiosity rover stumbled upon three large iron meteorites half-buried in the red sand.

## WATER WORLD

Billions of years ago, Mars had water flowing on its surface. New evidence from ground-based telescopes suggests that Mars even had an ocean that covered most of its northern hemisphere. Some of the water is still there, frozen in the north and south polar ice caps. Some of it is below the surface in underground springs. Some of it was lost into space when Mars's magnetic field disappeared and the planet was scoured by solar particles.

## PHOENIX LANDING

NASA's Phoenix spacecraft landed near the north pole of Mars in 2008 with a scooper and a mission: find water. The plan was to use Phoenix's robotic scoop to dig and dig until it found subsurface water ice. Just prior to landing, the retrorockets of Phoenix blew the top layer of Mars's soil away revealing . . . water ice. Mission accomplished! Phoenix still performed some scoops, and water seeped into the holes, froze, and then disappeared into the atmosphere.

## UNDERGROUND WATER EROSION

Water still shapes parts of Mars—from below. New gullies have formed from underground water eroding hillsides, and in 2015, scientists observed dark lines of material actually flowing downhill. They were carried by a salty brine of water that condensed from elements in the thin Martian air.

## CRATER LAKES

There are ice lakes at the bottom of some Martian craters. Frozen water has formed at the basins of these craters or seeped from below through surface cracks. One such ice lake located in an unnamed crater in the arctic circle of Mars is almost ten miles across and sports a shiny, blue-white exterior.

## MUSIC ON MARS

The first song broadcast from Mars was "Reach for the Stars," by will.i.am on August 28, 2012. NASA radioed the musical data 150 million miles from Earth to the Mars Curiosity rover, which in turn beamed it back to Earth and performed it before a live audience. Will.i.am, a huge NASA fan, wrote the song just for the occasion.

## MARTIAN METHANE

Mars has methane. On Earth, most of the methane is a result of biologic processes. Scientists go back and forth about what methane on Mars may mean. Are underground life forms creating methane and replenishing the atmosphere with this gas, or is there a natural, geological explanation? These outbursts of methane are happening currently, so if the cause is biological, something is alive under Mars.

# CHAPTER 6

# THE GAS GIANT PLANETS

*Jupiter, Saturn, Uranus, and Neptune*

Jupiter, Saturn, Uranus, and Neptune have a lot in common. They're all large (compared to Earth). They're all mostly made of gases. They all have rings. And aside from the Sun, they are the prime movers and shakers in the solar system. Jupiter vacuums up stray asteroids, diverts comets, and keeps a retinue of little objects nearby. Saturn holds many moons that create, mold, and shape its beautiful rings. Uranus and Neptune move each other and dominate the outer solar system with their strong gravitational pulls.

What do these four giants have in common with the four small, rocky planets (Mercury, Venus, Earth, and Mars) covered so far? Not much. Which means that you have a lot of weird worlds to explore. In this chapter you'll probe into the swirling surfaces of these gas giants, their multiple moons, and the environments they've carved out for themselves far from the Sun.

# JUPITER

## FAR-OUT JUPITER

Jupiter is the fifth planet from the Sun, but there is a huge gap between it and the fourth planet, Mars. While Mars is, on average, 140 million miles from the Sun, Jupiter orbits the Sun at a distance of 483 million miles.

## GIANT PLANET

Jupiter is just plain big. With an equatorial diameter of about 88,000 miles and a mass 318 times that of Earth, Jupiter is easily the largest planet—1,325 Earths could fit inside it. In fact, Jupiter is more massive than all of the planets, moons, and asteroids in the solar system combined.

## GREAT BALL OF GASES

Jupiter is not a solid planet like Earth or any of the inner planets. The vast majority of the giant planet is made of hydrogen gas mixed with a little helium. The surface that you see in pictures is only the upper layer of hydrogen clouds—Jupiter's ever-changing weather—which obscure a deep, dense layer of liquid and metallic hydrogen.

## NOT SO MUCH A SUN

Jupiter's composition is very similar to the Sun, and so you may have read on the Internet that "if only Jupiter were a little larger, it could be a star like the Sun." Actually, Jupiter would need to be about 100 times more massive in order to become just a small reddish star.

## WHAT'S THE FORECAST?

Unmanned spacecraft have relayed that the weather forecast on Jupiter can include violent storms, dramatic lightning, and a chance of auroras similar to those seen on Earth. This volatile weather happens because Jupiter's upper atmosphere is constantly churned up by internal heat and external interactions with the Sun.

## RAPID ROTATION

For a gigantic planet, Jupiter spins incredibly fast—a day goes by in just ten hours. Whirled around Jupiter by this speedy rotation, the clouds channel into horizontal bands that encircle the planet. Inside the broadest band of the southern hemisphere, you can glimpse Jupiter's distinguishing mark: the Great Red Spot.

## OFF THE CHARTS

Jupiter's famous Great Red Spot is a churning cyclone that rotates counterclockwise at more than 200 miles per hour. The Great Red Spot is so big two Earths could easily fit inside, which makes this one huge high-pressure system. And the Great Red Spot is no passing storm. The spot you see today most likely formed in the 1870s.

## THE OLD GREAT RED SPOT

Italian astronomer Giovanni Cassini documented a red spot on Jupiter between 1665 and 1712. But when the founder of the Cincinnati Observatory, Ormsby MacKnight Mitchel, observed Jupiter in the 1840s he saw many features including stripes, swirls, and dark and light patches, but never a Great Red Spot. What happened to the old spot? No one knows.

## RED JUNIOR

In 2006, a never-before-seen red spot popped up. Called Red Junior or the Little Red Spot, this feature arose from the merger of three white ovals in the South Temperate Belt (just south of the Great Red Spot).

## R.I.P. RED THE THIRD

In May 2009, observing teams for the Hubble and Keck telescopes discovered that a white oval had turned red in Jupiter's South Equatorial Belt. No one grew attached to "Red III," which was good because it only lasted a few months before it collided with the Great Red Spot and was obliterated.

## HEY, WHERE'D THAT STRIPE GO?

In 2010, a 250,000-mile-long stripe on Jupiter, the South Equatorial Belt that houses the Great Red Spot, suddenly disappeared from view. Most likely the stripe was still there but was obscured by a lighter-colored cloud layer covering it. In 2011, the stripe returned to prominence just as suddenly as it disappeared.

## JUPITER'S HIDDEN SECRETS

The only part of Jupiter astronomers can study in depth is that thin layer of atmosphere where the clouds condense. What lies below is still a mystery—as is why certain clouds visibly rise to the top while others sink out of view, or why the Great Red Spot is red (most of the time).

## IS IT REALLY RED?

Although the Great Red Spot has survived for more than a century, astronomers have observed some transformations. Most notably, its color varies from pale pink to bright red. Also, the Great Red Spot grows and shrinks when it interacts with other storms in the planet's atmosphere such as white and brown ovals.

## IT'S SHRINKING!

The shape of the Great Red Spot changes slowly over time. Over the last 100 years its characteristic oval shape has given way to a rounder appearance. If this trend continues, the Great Red Spot could be a much smaller, perfect circle soon—or it could disappear altogether.

## USE YOUR NAKED EYE

Jupiter is easily seen with the naked eye. Only the Moon, Venus, and Mars (on rare occasions) can be brighter than Jupiter in the night sky. Jupiter is about three times brighter than the brightest star, Sirius, and shines with a steady, cream-colored light.

## ANOTHER YEAR, ANOTHER CONSTELLATION

It takes Jupiter 11.86 Earth years to orbit the Sun. That means each year Jupiter shifts its position in the sky by $\frac{1}{12}$ of a circle. This circle is called the zodiac, and Jupiter visits each zodiac constellation in order, one by one, year after year. For instance, if you see Jupiter in front of the stars of Virgo, the next year it will be in front of Libra, then Scorpius, Sagittarius, and so on.

## BANDS OF LATITUDE

In a telescope Jupiter looks like a little disc with stripes on it. Depending on your telescope and the sky conditions, you can see two thick stripes bracketing the planet's equator and several thinner bands in the northern and southern hemispheres. You'll also see four "stars" lined up around Jupiter's equator. Jupiter has at least sixty-seven moons, but only the largest four can be seen in a backyard telescope.

## HIDE-AND-SEEK MOONS

Italian astronomer Galileo Galilei first charted Jupiter's four largest moons and their orbits in 1610. These Galilean moons are named (in order from closest to farthest from Jupiter) Io, Europa, Ganymede, and Callisto. Each night these four moons form a different pattern. One night there are two on one side and two on the other. Another night you might see three on one side and only one on the other. And sometimes you'll only see three moons because one is hiding in front of and behind the planet.

## BETTER THAN 20/20

Some rare humans possess unbelievably amazing eyesight and can see the four Galilean moons with their naked eye. While pointing out Jupiter to a class, a sharp-eyed first-grader asked, "What are those stars right next to Jupiter?" He continued, "There's like three to the left of Jupiter and one to the right." The telescope confirmed exactly what he saw with his naked eye.

## NOW ERUPTING

Jupiter's moon Io is the most volcanic place in the solar system. With volcanoes erupting daily, the topography of this world changes with frightening rapidity. Io builds mountains in a matter of months. While the Earth's Moon is mostly geologically dead, Io has volcanoes that shoot plumes of lava 250 miles above its surface. A caldera of molten magma on Io, named Loki, emits more heat than all the volcanoes on Earth combined.

## JUPITER'S TROJAN SQUAD

In addition to Jupiter's sixty-seven moons, it has a bunch of rocks that like to hang out with it. These objects, called trojans, circle the Sun with Jupiter—some of them go ahead of Jupiter and others trail behind its orbit. As of 2015, astronomers have found more than 6,000 trojans flying with Jupiter.

## COSMIC CARNAGE

Jupiter is so massive that objects are drawn to it, and astronomers have seen several objects hit it. In 1994, as predicted by astronomers, a comet named Shoemaker–Levy 9 first broke up into twenty-one pieces and then smacked into Jupiter's atmosphere. The Hubble Space Telescope imaged the carnage—a series of black burn marks the size of Earth that took Jupiter's atmosphere months to obliterate. Meteors or comets also hit Jupiter in 2009, 2010, 2012, and 2016.

## LOOKING BACK ON THE RINGS

Jupiter has a small ring system that is almost impossible to see from Earth. Astronomers only found these rings when the Voyager 1 spacecraft flew past Jupiter in 1979, turned around, and took some pictures of Jupiter silhouetted by the light of the Sun. And there they were, these oh-so-faint rings that are mostly small particles of dust. The Halo ring is closest to the planet followed by the Main ring, the Amalthea Gossamer ring, and the Thebe Gossamer ring.

## FLYING BY JUPITER

Seven spacecraft have flown by Jupiter: Pioneers 10 and 11, Voyagers 1 and 2, Ulysses (on a long, round-about way to the Sun), Cassini (on its way to Saturn), and New Horizons (on its way to Pluto).

## STAY AWHILE

The Galileo mission orbited Jupiter from 1995 to 2003 and relayed close-up images and data. Scientists ended the mission by crashing Galileo into Jupiter in order to learn more about the Jovian atmosphere and to avoid accidentally contaminating one of Jupiter's moons with Earth microbes that may have hitched a ride.

## BACK IN BUSINESS

Between 2003 and 2015 there weren't any spacecraft in orbit around Jupiter. That monitoring drought ended when NASA's Juno mission arrived in the summer of 2016. Juno will try to pick up where Galileo left off and provide updates on the changing surface of Jupiter and the dynamic landscapes of its moons.

## KING OF THE PLANETS

Across cultures, the planet Jupiter was most often associated with the king of the gods. Whether you call it Jupiter, Jove, Zeus, Thor, Amun, or Indra, Jupiter is definitely king.

## WHAT LURKS BELOW?

Although Europa looks like a giant ice ball, deep down it is not as cold as it looks: Europa is frozen, but not frozen solid. Europa's smooth, icy surface is broken by stress cracks and shattered sporadically by meteor impacts. And, something is sloshing around below the ice—something liquid—that astronomers think is $H_2O$. In fact, there may be more water on Europa than on Earth. Where there is water on Earth, there is life. Could there be aquatic life in Europa? Visiting this moon of Jupiter is a high priority for NASA.

# SATURN: RING WORLD

## IS THAT GAS?

Saturn is a gas giant planet that is 75,000 miles in diameter and composed of hydrogen and helium. Saturn spins really fast for being such a large planet. A day on Saturn is only 10.5 hours long. However, it is much farther from the Sun and so its year is the equivalent of 29.5 Earth years.

## LONG-DISTANCE LOVE

Saturn is the sixth planet from the Sun and circles the Sun at an average distance of 888 million miles. When Saturn is closest to Earth, it is still about 750 million miles away.

## BATTLE OF THE BULGE

Saturn is the flattest—or most oblate—planet in the solar system. In a telescope, it looks like a flattened oval. Saturn's low density plus high rotation rate has made it decidedly unround. It is squished at the poles and bulges at its equator more than any other planet.

## THE HEXAGONAL HURRICANE

A raging hurricane the shape of a hexagon 20,000 miles wide swirls on the north pole of Saturn. The hexagon is an atmospheric disturbance caused by jets of gas swirling below the surface.

## LET ME FLOAT THIS BY YOU

Saturn, as a whole, is less dense than water. That means that if you built a bathtub large enough to hold Saturn and filled it with water, Saturn would float. And, of course, it would probably leave a ring!

## BETTER PUT A RING ON IT

Saturn's rings are composed of millions of separate particles of ice and dust that reflect the light of the Sun. And each ring orbits Saturn at a different rate—the inner rings orbit faster than the outer rings. Saturn's main rings stretch more than 200,000 miles in length, yet in some places are only a fragile thirty-feet thick. The origins of the rings are still a mystery, but evidence suggests that they were formed when several moons either collided above Saturn's atmosphere or were shredded by Saturn's intense gravity.

## LOOK AT THOSE . . . EARS?

When Galileo first spied Saturn and its rings in his telescope in 1610, he didn't know what he was seeing. He explained his blurry view as best he could when he wrote, "Saturn has ears." Dutch astronomer Christiaan Huygens had a much better telescope, and in 1655 he proposed what he was seeing around Saturn was not ears but a thin, flat ring. In 1675, Huygens's contemporary, Giovanni Cassini, observed a wide, black gap in the rings through his telescope. That area, which is largely free of ring particles, is still called the Cassini Division.

## SHEPHERD MOONS

Several small satellites orbit within or near Saturn's distinctive rings. Astronomers believe that these "shepherd moons" play a vital role in maintaining the size and shape of the rings. They act as gravitational tugboats that keep the rings from flying into space or crashing into the planet below.

## MAKING WAVES

Shepherd moons leave their marks on the ring material. The moons push their weight around and even create waves in the ring particles. And some of the ring material clumps into movable mountain ranges two miles high. One tiny, five-mile-wide moon named Daphnis was nick-named the "wave maker" since it left a visible gravitational wake in the rings that bracket its orbit.

## MOONS IN THE MIDDLE

Saturn's rings have large, dark emptier areas called gaps or divisions. And some gaps have moons in them. What came first, the gaps in the rings or the moons? Did the moons make the gaps, or did the material, once in the gaps, make the moons? Astronomers are not sure.

## PRIMITIVE EARTH

Titan is Saturn's largest moon, second-largest moon in the solar system, and a little larger than the planet Mercury. Astronomers think Titan is a world that may have many elements in common with an early Earth. But don't pack your bags for Titan just yet. You couldn't breathe the air on Titan. It's over 98 percent nitrogen, and the rest is mostly methane. If you like oxygen, Titan is not the place for you. And it's pretty cold—like −290°F.

## SAILING THE SEAS OF TITAN

Despite being incredibly cold, Titan has flowing liquid on its surface. You wouldn't want to drink it or skinny-dip in it, however. The most common liquid is not water but methane. On Earth methane is a gas, but with such cold temperatures on Titan, it takes the form of a liquid. There are methane lakes and seas on Titan that erode the shores and exhibit tides. And within these seas are islands of land sticking above the waves.

## EARTHLY VISITATION

Saturn has had four earthly visitors. Three unmanned spacecraft flew by the ringed planet (Pioneer 11 in 1979, Voyager 1 in 1980, and Voyager 2 in 1981), and one entered orbit in 2004 (Cassini).

## THIRTEEN LUCKY YEARS

It took almost seven years for the Cassini spacecraft to reach Saturn. For more than a decade it has closely monitored changes in Saturn's rings, weather, and moons, and has sent back breathtaking images and a wealth of data. In 2017, after thirteen years in orbit and with its fuel supplies running low, Cassini will purposely plunge into Saturn's atmosphere.

## SATURN SEASON

You can see Saturn with the naked eye. It looks like a semibright yellow star. Saturn is at its biggest and brightest in the nighttime sky every fifty-four weeks when it is closest to Earth.

## THE PLANET THAT LOITERS

Since Saturn takes 29.5 Earth years to orbit the Sun, it moves very slowly across the background stars that you can see from Earth. From week to week, month to month, Saturn will barely move with respect to the zodiac constellations. In fact, Saturn will hang out with each constellation for about two and a half years.

## A TELESCOPIC VIEW

What you see when you look at Saturn through a telescope depends on the size of the telescope. Small telescopes will pique your interest with a tiny oval-shaped light. Moderate telescopes will make you gasp when you see a ring around such an improbable planet. And large telescopes reveal a wonderful world with multiple rings, ring shadows on the planet, and planetary shadows cast on the rings. You'll say, "Wow!" Guaranteed.

## FADING GLORY

When you observe Saturn in a telescope you may or may not see colors on the planet. It depends on your eyes. Younger eyes see yellow, orange, and green on Saturn, while older eyes see only white. Another reason to look through a telescope this year—your eyes aren't getting any younger!

## MOON VIEW

The number of moons you can see around Saturn depends on the telescope you use. The smaller scopes will reveal one to four moons (in order of brightness: Titan, Rhea, Tethys, and Dione). And with a larger scope you can add three more moons to your viewing experience (Iapetus, Enceladus, and Mimas).

## SATURNIAN SEASONS

Saturn is tilted 27 degrees with respect to its orbit around the Sun. This gives it seasons similar to Earth, but each season lasts almost thirty times longer. This slant changes your viewing perspective on the rings. When the rings are tipped up or down to you, Saturn reveals the classic views you're used to. That corresponds to Saturnian summer and winter. When the rings close down and become edge on to you, that marks spring and fall.

## NAKED SATURN

In some places Saturn's rings are barely 30-feet deep. That means when the plane of the rings are pointing directly at Earth (as they were in September 2009), they are invisible through even the most powerful telescopes. When Saturn seems to be missing its rings, astronomers sometimes call this "naked" Saturn.

# URANUS: BEST PLANET TO SAY ALOUD

## SAY WHAT?

There are several customary ways to pronounce the word *Uranus*. Both "YOOR-ah-nus" and "your-AY-nus" are acceptable. Astronomers tend to vary their pronunciation based on the maturity level of their audience.

## THE MADNESS OF NAMING IT GEORGE

Uranus was the first planet discovered by a telescope. English astronomer William Herschel found Uranus in 1781. Herschel wanted to name it after the king of England, George III. Other astronomers wanted to name it after its discoverer and called it "Herschel." Eventually a more traditional name from ancient mythology, Uranus, became the standard.

## ONLY GREEK PLANET

While every other planet in the solar system is named for a god in Roman mythology, Uranus comes from the creation myths of the ancient Greeks.

## ANCIENT GREEK SOAP OPERA

Uranus (the god) was the father of Cronus (Saturn in Roman mythology). Cronus, who hated his dad, castrated Uranus with a giant, jagged sickle fashioned by his mom-god Gaia. Cronus then threw the discarded member into the sea where some of the semen magically turned into Aphrodite, goddess of beauty. What a crazy myth. Thanks, ancient Greeks!

## TESTING THE LIMITS OF YOUR EYESIGHT

Technically Uranus is bright enough to be seen with the naked eye. But you need to know exactly where to look, have perfectly clear skies, amazing eyesight, and almost no light pollution.

## DOUBLE THE DISTANCE

Uranus is the seventh planet from the Sun and lies about 1.8 billion miles away from the Sun (about twice as far as Saturn is from the Sun). A day on Uranus is seventeen hours long, but it takes Uranus eighty-four Earth years to orbit the Sun. Since its discovery in 1781, astronomers have observed it for only two complete trips around the Sun.

## A WORLD ON ITS SIDE

Uranus's main claim to fame is that it is tilted on its side compared to its path around the Sun. While Earth is titled 23.5 degrees, Uranus is tipped over almost 98 degrees. This makes days, nights, and seasons really complicated.

## FROM ONE EXTREME TO ANOTHER

For twenty-one Earth years, one hemisphere faces the Sun while the other hemisphere is always in the dark. Then during the twenty-one-year spring, when the planet looks like a bowling ball rolling along its orbit, everywhere gets equal hours of day and night. Then the other side of Uranus gets twenty-one years of sunlight while the hemisphere that got so much light before languishes in darkness. Twenty-one years of fall (and equal day and night) follow, and then the cycle repeats itself.

## MAKING EARTH LOOK SMALL

Compared to Earth, Uranus has four times the diameter, sixty-three times the volume, and more than fourteen times the mass of our planet.

## A LIGHTWEIGHT AMONG GIANTS

Although Uranus is much bigger than Earth, it is not in the same weight class as Jupiter or Saturn. It's less than half the diameter of Saturn and a little more than 1/3 the diameter of Jupiter. Furthermore, Saturn is almost five times more massive, and Jupiter is more than twenty times more massive than Uranus.

## LIGHTEST OF THE BIG PLANETS

Like Jupiter and Saturn, you can't stand on Uranus. It is made almost entirely of hydrogen and helium gases. The ratio of these gases makes Uranus the least massive of the gas giant planets. Although it's slightly larger than Neptune, Uranus is much less dense.

## BULLS-EYE!

Uranus has the second-largest and brightest rings in the solar system. One extra cool thing with Uranus is that since it is tipped so dramatically, sometimes astronomers can see all the way around the rings. When the planet sits in the center of the rings like a bulls-eye, these complete circles of material give Uranus a stunning halo.

## WITH A LITTLE MAGNIFICATION

You can find Uranus with even a small pair of binoculars. It looks like a pale blue dot. In a telescope it looks like a pale blue disc. In fact, it was this disc shape that got William Herschel's attention when he discovered it. Uranus did not look like a pinpoint—like a star—but was round like the other planets.

## ALMOST FEATURELESS

Unlike the other gas giant planets, Uranus shows very few surface features. It has storms and changing weather patterns, but a hazy green-blue layer of clouds seems to always cover the outer surface of the planet. Below are churning cyclones of gas that only rarely break the surface.

## WHERE ARE THE MOONS?

Uranus has twenty-seven known moons. You'd need a very large telescope to see any of them for yourself. The moons Titania and Oberon are the brightest, but even they look like tiny dots in a large scope.

## A MOON BY ANY OTHER NAME

Where have you heard the names Titania and Oberon? These and other names for Uranus's moons, like Miranda, Ophelia, and Juliet, are also characters from the works of William Shakespeare. Sticking with the Shakespearean theme, Oberon has a crater over 100 miles across called Hamlet. Other moons of Uranus, like Umbriel, Ariel, and Belinda, come from the poem "The Rape of the Lock" by Alexander Pope, a seventeenth-century English poet.

## LONG WAY DOWN

Uranus's moon Miranda shows some seriously extreme geologic processes. Instead of volcanoes of lava, Miranda grew mountains of ice that erupted out of the interior like white daggers. One of these daggers on Miranda is the largest cliff in the solar system: Verona Rupes. With an estimated vertical drop of more than six miles the top of Verona Rupes would make an amazing sight.

## LONE VISITOR

Voyager 2 is the only spacecraft that has explored Uranus at close range, when it flew within 51,000 miles of the pale blue planet in 1986. Although other missions have been proposed to revisit and possibly orbit Uranus, none have been funded.

# NEPTUNE: THE FINAL PLANET (FOR NOW)

## JUST. SO. SLOW.

Neptune is the eighth and farthest planet from the Sun, and it inhabits an orbit almost 3 billion miles from the Sun. It is the slowest-moving planet and takes 164 Earth years to orbit the Sun once.

## NO-STANDING ZONE

Like the other gassy planets, you can't stand on Neptune. It's made of a similar combination of hydrogen, helium, and other gases. But if you could visit the surface of Neptune, the force of gravity would be the second strongest of the planets in the solar system, after Jupiter.

## DIFFERENTIAL ROTATION

On average, Neptune's day is sixteen hours and six minutes long, the slowest rate of the four gas giants. Since it is made up of mostly gases, different parts of the planet spin at different rates. The equator takes about eighteen hours to spin once, while the surface surrounding the poles takes only twelve hours to rotate.

## IT'S SO DENSE

Although Uranus is slightly larger than Neptune, Neptune is slightly more massive. Why's that? Because its gases are denser than those of Uranus, Jupiter, or Saturn.

## NO COMPARISON

Neptune and Earth are the fourth- and fifth-largest planets in the solar system, respectively. However, it is a humongous jump from fifth to fourth. Neptune is almost four times as wide and over seventeen times as massive as Earth.

## WATERY NAMES

Neptune has fourteen known moons. In keeping with Neptune's nautical name, these moons bear the handles of minor water deities from Greek mythology, like Triton (Neptune's largest moon), Naiad, and Nereid.

## THERE'S A RIGHT WAY AND A WRONG WAY

Neptune's largest moon, Triton, is definitely an oddball. It is the largest moon in the solar system to revolve around its planet in the wrong direction. What's the "right" direction? Most moons orbit a planet in the same direction that the planet spins. But as Neptune rotates rapidly in one direction every sixteen hours, Triton orbits in the opposite direction every 5.88 days. This peculiar behavior suggests to astronomers that Triton was not formed with the planet Neptune but was captured billions of years ago by the big blue planet's gravity.

## DIMMING RINGS

The five rings of Neptune are extremely dark, but astronomers found them in a tricky way. When the faint rings passed in front of backgrounds stars, they partially eclipsed the distant starlight. Astronomers could measure this slight dimming and map the extent of Neptune's rings.

## WHO NEEDS A TELESCOPE?

Neptune was discovered in 1846 by two guys who didn't even look for it in a telescope. John Couch Adams in England and Urbain Le Verrier in France independently found this planet on paper—using math. Le Verrier was so assured in his discovery that he didn't even need to see it for himself. He instructed Berlin Observatory astronomer Johann Galle to aim his telescope at the coordinates he provided. Galle did so and found Neptune only 1 degree from where Le Verrier said it was.

## PLANET-NAMING POLITICS

Many people (especially in France) wanted to name the new planet after Le Verrier. They tried smoothing over their rivalry with England by agreeing to call Uranus "Herschel's Star" if the English would call the new planet "Le Verrier's Star." But in the end, the international community of astronomers declared "Neptune" to be the official name of this planet.

## GAS GIANT SUBSETS

Lately astronomers have referred to Uranus and Neptune as "ice giants" to differentiate them from their larger cousin-planets, Jupiter and Saturn. When looking outside of the solar system, they've found similar types of planets: gas giants like Jupiter, ice giants like Neptune, and rocky planets like Earth.

## COMPLETING THE GRAND TOUR

Like Uranus, only the Voyager 2 spacecraft has flown close to Neptune. Voyager 2 left Earth in 1977 and zoomed past Neptune in 1989.

## DEEP BLUE DOT

Since Neptune is so far away, it is not visible to the naked eye. In fact, you would need a good pair of binoculars or a small telescope just to find it as a tiny dot. Only in a larger telescope could you resolve the disc of the planet, where it shines with a nice deep blue color.

## THAT'S NOT WATER

It may look like water, but the surface of Neptune is colored this shade of blue because of a little extra methane gas in its atmosphere. You definitely would not want to swim in the methane-rich atmosphere of Neptune because it would be the last swim you ever took.

## THE GREAT DARK SPOT

Compared to the seemingly featureless surface of Uranus, you can see a lot more weather happening on Neptune. High, light-colored clouds often encircle the planet, and dark storms rage. Remember the Great Red Spot on Jupiter? Well, Neptune has a similar storm on its surface the size of Earth that is dark in color. Guess what they call it? The Great Dark Spot.

## A SPOTTY SPOT

Neptune's Great Dark Spot was "spotted" by the Voyager 2 spacecraft in 1989, but by the time the Hubble telescope looked at Neptune in the 1990s, no dark spot was spotted. Where did it go? Another dark spot appeared in a different location. Where'd that one come from? Good questions. With 3 billion miles between Neptune and the astronomers, it is tough to find concrete answers until there are better telescopes or more close flybys.

## HOLD ON TO YOUR HATS!

Neptune is the windiest planet. Astronomers clocked sustained winds of 1,300
miles per hour on its blue and white
cloud tops.

## SCOOTING WITH THE DARK SPOT

"Scooter" was the nickname given to a white storm on Neptune that
moved faster than other features on the rapidly rotating planet. You can
see Scooter cozying up to the Great Dark Spot in many of those classic
pictures of Neptune taken by the Voyager 2 spacecraft.

# CHAPTER 7

# PLUTO AND OTHER SMALL STUFF

*Weird, Wild Worlds*

When you were a kid, Pluto was a planet. At the time, astronomers knew of nine round objects that went around the Sun (the planets), and Pluto was always the oddball. It was the smallest, the coldest, the farthest, and the planet with the weirdest orbit. But it was always a planet, all the more likeable because of its eccentricities.

However, in the twenty-first century, astronomers grew up. They put aside their smaller telescopes of the past and utilized the latest technology. With these updated telescopes, they learned that the solar system was much, much more complex. And they learned that Pluto, it turned out, was not alone, and was not even unique. It had more than 1,000 brother, sister, and twin worlds way out there. These new objects showed that Pluto didn't fit into the categories astronomers had previously set up.

If you grew up thinking Pluto was a planet, it's time for you to come of age. Pluto isn't a planet—and it isn't weird at all, but that's not saying that there isn't some weirdness out there. In this chapter you'll learn about asteroids, comets, and other oddities of the solar system, like the wild, wonderful world called Sedna. Once you jettison your childish notions of Pluto, you may see that Sedna is the new Pluto!

# PLUTO AND
# ITS NEW FAMILY

### ONE QUICK TRIP

Pluto is a spherical ball of ice and rock 1,474 miles in diameter. That's about the distance from Washington, D.C., to Denver, Colorado.

### A LONG AND VARIED YEAR

It takes Pluto 248 Earth years to travel once around the Sun, but its distance to the Sun varies quite a bit. It can be as close at 2.75 billion miles from the Sun and as far as 4.58 billion miles. In fact, for a portion of its orbit, Pluto is closer to the Sun than the planet Neptune is.

### SAME PLACE, DIFFERENT TIME

Although it would be unbelievably cool, Pluto and Neptune will never run into each other. In three-dimensional space, their orbits don't overlap and never cross paths. And almost every time Pluto is closest to the Sun, Neptune is more than 3 billion miles away from it.

### STRANGELY BRIGHT

In 1930, American astronomer Clyde Tombaugh discovered Pluto. In spite of its distance from Earth, Pluto seemed very bright. Because of its relative brilliance, astronomers originally thought it nearly the size of Earth. Upon further investigation, Pluto was revealed to be covered by a highly reflective icy coating that made it look bigger than it actually was. It is not a rocky object like Earth and therefore isn't very dense. In fact, Earth is 500 times more massive than Pluto.

## MOON DANCE

Pluto has five moons: Charon, Nix, Hydra, Styx, and Kerberos. Charon, Pluto's largest moon, is about ⅔ the diameter of Pluto, which causes an interesting dance. Charon is so massive that it does not simply orbit Pluto like most moons. Instead, the two orbit a point in between them (albeit closer to Pluto than Charon), which means that Pluto and Charon pretty much orbit each other.

## ALWAYS WATCHING

Pluto and its largest moon, Charon, face each other all the time. While Pluto rotates once every six days, Charon revolves around Pluto once every six days. So one hemisphere of Pluto never sees Charon and one side of Charon never sees Pluto.

## WHAT IS A PLANET?

Here's the official definition from the International Astronomical Union (IAU), the "deciders" in the astronomy world: a body that orbits the Sun, is large enough to be round, and is massive enough to have cleared its orbit around the Sun. The intent of the definition is that, after the Sun, the planets are the next uniquely important category of objects in the solar system. They are the movers and shakers. They dominate anything else that comes nearby.

## PUSHED AROUND BY NEPTUNE

Pluto orbits where it does because Neptune allows it. If things were different, Pluto would be perturbed by Neptune's immense gravity and flung around the solar system. However, Pluto is now tucked in behind Neptune's wake and has found a stable place to continue orbiting the Sun. Neptune, as the mover, is a planet. Pluto, as the move-ee, is not.

## IT'S A WILD WORLD

Astronomers know a lot about Pluto thanks to the New Horizons spacecraft, which flew by the tiny world in 2015. The close-up pictures taken by New Horizons are incredible. They show that Pluto, small as it is, has an atmosphere. There are high mountains, flat plains, and even signs of ice volcanoes on its surface.

## GETTING THERE

New Horizons was the fastest spacecraft ever launched from Earth. However, Pluto is so far away that it still took nine and a half years to get there.

## NEW HORIZONS PART TWO

The New Horizons spacecraft, which flew by Pluto in 2015, is on its way to rendezvous with another object even farther from the Sun: a tiny ball of ice thirty miles across called 2014 MU69. This ice chunk seems to be part of a population of objects very far from the Sun that astronomers call Kuiper belt objects, trans-Neptunian objects, or plutoids. The plucky little spacecraft is due to visit 2014 MU69 on January 1, 2019. By then, astronomers should have come up with a better name for it.

## THE FIVE DWARFS

There are five official dwarf planets in the solar system: Pluto, Eris, Haumea, Makemake, and Ceres. According to the IAU, a dwarf planet is a celestial body that orbits the Sun, is large enough to be nearly round, is not a satellite of another object, and has not cleared the neighborhood around its orbit, which means that it shares its orbit with other similar objects like asteroids and plutoids.

## WARRIOR PRINCESS

Eris is another round dwarf planet that lives in the far reaches of the solar system. Lying more than 6 billion miles from the Sun, Eris was first seen in 2005 by American astronomers Mike Brown, Chad Trujillo, and David Rabinowitz. Eris's discoverers originally nicknamed it Xena, from the TV series *Xena: Warrior Princess*. The team leaked the tongue-in-cheek homage to a 1990s television star to the media, but the dwarf planet was soon given its more appropriate, official name of Eris, from the goddess of discord in Greek mythology.

## LUCY LAWLESSNESS

Eris has one moon, which was named Dysnomia after an obscure Greek myth related to Eris. *Dysnomia* means "lawlessness" in ancient Greek. According to Mike Brown, the name's connection to Lucy Lawless, the actress who played Xena, was pure coincidence.

## THE TENTH PLANET?

When Eris was first discovered, some astronomers considered it to be the tenth planet. However, other astronomers discovered so many other objects in the far reaches of the solar system in the 1990s and 2000s that in 2006 they voted to reclassify them all. If Eris was not discovered, Pluto might still be a planet.

## FIRST RUNNER-UP

Until 2015, astronomers believed that Eris was larger than Pluto. That was until New Horizons measured Pluto accurately. At 1,445 miles in diameter, Eris is not quite as big as Pluto (which is 1,474 miles across). However, Eris still outweighs Pluto since it is made of denser material.

## MIRROR, MIRROR

The surface of Eris is covered in white ice and is more reflective than any planet or dwarf planet. It's almost like a mirror. This made Eris look relatively bright in a telescope and led astronomers to believe that it was a huge object. But Eris isn't bright because it's big. It's bright because it's superb at reflecting weak rays of the Sun.

## ARE YOU READY FOR SOME FOOTBALL?

Another dwarf planet, named Haumea, is a crazy football-shaped object that rotates on its axis every four hours. Two small moons, named Hi'iaka and Namaka, and a debris field accompany it on its journey around the Sun. Something smacked into Haumea a long time ago leaving the mis-shapen mass observable today.

## A MATE FOR MAKEMAKE

Makemake is another dwarf planet in the far-flung region of the solar system. Astronomers know it is there because it showed up as a tiny, slow-moving point of light during a sky survey in 2005 by Mike Brown's team. But not a lot is known about it since it is almost 5 billion miles from Earth and is smaller than Pluto. However, in 2016 astronomers announced that Makemake has at least one moon that's about 100 miles wide, but we still have a lot to learn.

# SEDNA: A NEW ODDBALL WORLD

## UNIDENTIFIED FLYING SEDNA

In 2003, astronomers discovered a previously unidentified chunk of ice in the outer solar system that they named Sedna. This round object about ⅔ the size of Pluto is neither a planet nor a dwarf planet. Astronomers are not quite sure what to make of it and why it is there.

## COULD IT BE A COMET?

Sedna has a highly elliptical orbit. At its closest, Sedna is still over 7 billion miles from the Sun. But when it is farthest from the Sun it will be about 87 billion miles away. Sedna's orbit looks more similar to that of a comet than a planet.

## HEY, WHERE'RE YOU GOING?

Since Sedna is so far away it takes it about 11,400 Earth years to go around the Sun one time. Other than miniscule, long-period comets, it may eventually be the most distant object in the solar system.

## ROGUE PLANET

Astronomers don't know how to categorize Sedna. Some consider it to be a scattered-disc object, something thrown out there by Neptune's gravity. Others think it is a member of the Oort cloud, an elusive population of comets that can only be seen when they stray closer to the Sun. Maybe a passing star moved Sedna to its strange location. Maybe an unseen, large planet lies farther away in the Oort cloud. Or perhaps Sedna is a rogue planet that migrated from an alien star system.

## FINAL FRONTIER FORERUNNER?

Sedna makes astronomers wonder whether there could be something else out there, something farther away and/or less reflective. Astronomers only discovered Sedna when it was relatively close to Earth, but it spends 90 percent of its life at a distance where it was nearly impossible to find. Could Sedna be the forerunner of another class of objects that spend the majority of their orbits at the edge of the solar system? If so, what do you call them? Planets? Dwarf planets? Other? That's why Sedna is so intriguing. It opens up so many possibilities.

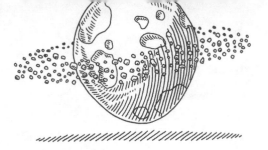

# ASTEROIDS

## ASTEROIDS ARE EVERYWHERE!

*Asteroid* is a broad term encompassing most of the rocky material that is not a planet or a moon. Asteroids range in size from microscopic to more than 500 miles in diameter and are generally irregular in shape. The majority of asteroids lie in a belt between the orbits of Mars and Jupiter. However, there is also a category of asteroids, called near-Earth asteroids, that lie inside Mars's orbit. And there are other asteroids that orbit the Sun out by Jupiter, Uranus, and Neptune.

## FINDING MORE THAN ONE ASTEROID EACH DAY

Astronomers have discovered more than 600,000 asteroids in the solar system, most of these in the past ten years. In fact, they are finding more of these leftovers of the solar system almost daily.

## SMALL, DISTANT, AND DIM

The materials that make up asteroids do not reflect much light. Since asteroids are so small and so far away, they are not very bright. In fact, you need binoculars or a telescope to see even the brightest asteroids from Earth.

## THE FIRST AND BIGGEST!

Italian astronomer Giuseppe Piazzi discovered the first asteroid on January 1, 1801. He named it Ceres after the ancient Roman harvest goddess. At 578 miles in diameter, Ceres is the largest asteroid and one of the few that is spherical in shape. That said, at only ¼ the diameter of the Moon, Ceres is still small in the grand scheme of things.

## HEAVY HITTER

In addition to being an asteroid, Ceres is also considered a dwarf planet because it is massive enough to achieve a nearly round shape. If you add up the masses of all the known asteroids in the solar system, Ceres alone would account for about 25 percent of the total.

## A ROUNDISH WORLD

Vesta is the brightest asteroid that is visible from Earth. It is the only asteroid that is technically visible to the naked eye, but conditions (and your eyesight) need to be superb. Vesta is about 326 miles in diameter and is *almost* round in shape, which unfortunately disqualifies it as a dwarf planet. It's round, but not round enough.

## TOWERING RHEASILVIA

The tallest mountain in the solar system is on minuscule Vesta. Rising fourteen miles above the surrounding plain, Rheasilvia is just slightly taller than the previous record holder, Olympus Mons, the largest mountain on Mars.

## THE PLANET THAT NEVER FORMED

Where did the hundreds of thousands of asteroids located in the asteroid belt between Mars and Jupiter come from? The leading theory is that most of the asteroids are the remnants of a small planet that never formed. Jupiter's massive presence in the area may have disrupted the planet-forming process long ago and kept them scattered into a widespread belt of rocks.

## ADDING UP THE ASTEROIDS

If the asteroids had come together to form a planet, it would not have been very big. If you add up all the 600,000 known asteroids and smash them together, their combined mass would be less than that of the Moon.

## NOT-SO-DANGEROUS DISTANCE

In the 1980 Star Wars film **The Empire Strikes Back**, Han Solo weaves his spacecraft the **Millennium Falcon** through an asteroid belt. That was great cinematography, but the asteroid belt in the solar system is nowhere near that dense. Most asteroids are thousands to millions of miles apart—so far apart that none of the spacecraft that have been sent through the belt to get to Jupiter, Saturn, and the outer planets have been damaged by an unseen rock.

## SPACECRAFT RSVP

The Galileo spacecraft, on its way to Jupiter, purposely made a close pass of three asteroids. It passed within 1,000 miles of Gaspra, an irregular-shaped asteroid twelve miles long and seven miles wide, in 1991. This was the closest pass any spacecraft had made to this type of object, and Galileo transmitted the first detailed images of an asteroid.

## SURPRISE!

In 1993, Galileo swung by an asteroid named Ida and astronomers were surprised to see that Ida had a moon. A one-mile-diameter asteroid, later named Dactyl, orbits about sixty-two miles above Ida's center of gravity. This was the first visual evidence of an asteroid having its own moon.

## THREE'S COMPANY

The asteroid named Sylvia has two small moons, named Remus and Romulus. Sylvia is about 175 miles in diameter, with four-mile-wide Remus and eleven-mile-wide Romulus circling it. This was the first triple-asteroid system discovered.

## ROMANTIC RENDEZVOUS

The NEAR (Near Earth Asteroid Rendezvous) Shoemaker spacecraft orbited the asteroid Eros 230 times and sent back thousands of close-up pictures. As a grand finale, NEAR actually landed on the surface of Eros in 2001 and beamed back information to awaiting scientists. Ironically, astronomers announced the successful rendezvous on February 14—a fitting date to meet with the god of love.

## SPACE POTATO

Eros is a good example of an asteroid. It is oddly shaped (like a potato twenty-one miles long, eight miles wide, and eight miles thick). The gray surface is dotted with craters from ancient impacts. Eros rotates every five hours and revolves around the Sun in 1.76 Earth years. Although Eros has been as close as 14 million miles from Earth, it is no threat to hit us.

## NOTHING BUT NET

Eros does have gravity—very weak gravity. A person weighing 200 pounds on Earth would only weigh 2 ounces on Eros. A basketball player with a thirty-six-inch vertical leap could jump one mile off the surface.

## COLLISION COURSE?

Astronomers have discovered more than 11,000 near-Earth asteroids. Only a few of these are causes of concern. Even the most troublesome, those classified as Potentially Hazardous Asteroids (PHA), have almost no chance of hitting Earth.

## CHOOSE WISELY

When a new asteroid is discovered, it is given a number. Then the discoverer of the asteroid suggests a name to the International Astronomical Union. Almost every suggested name is approved, but as their rules state, "names of pet animals are discouraged." No asteroid for Fluffy.

## FAMOUS NAMES

Many asteroids are named after famous people. For instance, there is 2620 Santana (for Carlos Santana), 16155 Buddy (for Buddy Holly), and 249516 Aretha (for Aretha Franklin). There's even an asteroid named 8815 Deanregas. You can check out the list of asteroid names from the Minor Planet Center, run by the Smithsonian Astrophysical Observatory at Harvard. With more than 600,000 of them and counting, you may find that there's an asteroid named after you.

## ASTEROID B-612

*The Little Prince* by Antoine de Saint-Exupéry is a story of a boy who travels from his home on Asteroid B-612 to Earth and discovers the joys and frustrations of life on a large planet. There is an asteroid 612 Veronika, and author Saint-Exupéry himself is memorialized in space as asteroid 2578 Saint-Exupéry.

## SMALL IN SPACE

A meteoroid is a small asteroid, but the terms are often used interchangeably. These stony or metallic objects can come from the asteroid belt, the Moon, and even other planets.

## ONE SHINING MOMENT

A meteor exists for only a brief moment. When a meteoroid, asteroid, or comet plunges through Earth's atmosphere and creates a "shooting star," *that* is a meteor. You can wish upon it, but you have to be quick. Most meteors are the size of a grain of sand and rapidly burn up before hitting the ground.

## COMET SHOWERS

Annual meteor showers like the Perseids and Leonids are caused by comets. A comet goes by, leaves parts of its tail behind, and then Earth runs into that debris. Like clockwork, on the same week every year, Earth flies through more leftover comet parts. The Perseids occur around August 12 each year and come from the tail of Comet Swift–Tuttle. The Leonids peak around November 17 and originate from the debris of Comet Tempel–Tuttle.

## AN ASTEROID SHOWER

The meteors from the annual Geminid meteor shower are made of asteroid parts. Asteroid 3200 Phaethon seems to be the culprit. As it regularly flies close to Earth, 3200 Phaethon leaves bits of rock and ice behind. And around every December 13, Earth flies through these asteroid parts and gets a solid meteor shower.

## FIERY FIREBALLS

Really bright meteors are called fireballs. They can glow all sorts of colors including white, blue, and green. Really dramatic fireballs can break up into multiple pieces and become multiple meteors streaking across the sky.

## THE GLOW IS GONE

When you see a meteor, it may look close to you but it's not. You see them shine when they are forty to fifty miles above Earth in the upper atmosphere. Meteors blaze because they are decelerating from tens of thousands of miles per hour to hundreds of miles per hour. That rapid deceleration transmits into heat and causes the air around the meteor to glow. Once the meteor is lower in the sky and has stopped this rapid deceleration, it stops shining.

## UNEXPECTED SONIC SOUND WAVE

On February 15, 2013, a huge meteor streaked across the sky over Chelyabinsk, Russia. For a brief moment it shone brighter than the Sun and cast stark shadows. At first it did not make a sound, but about two minutes later a sonic boom shattered windows and even knocked people over. The speed of sound is 767 miles per hour—much slower than the speed of light at 186,000 miles per second, which accounts for the lag time between seeing it and hearing it.

## METEORIC FALL

Astronomers think the phrase "meteoric rise" is hilarious since, in their field of study, meteors only fall, burn up, quickly fade, and, occasionally, unceremoniously smack into the ground.

## SPACE ROCKS ON EARTH

When a meteor survives its fiery plunge through the atmosphere and hits the ground, that object is called a meteorite. These objects are asteroids or meteoroids when they are in space, meteors when they are briefly falling through the atmosphere and streaking across the sky, and meteorites when you can pick them up and hold them.

## ROCKY ORIGINS

Most meteorites originate from asteroids or meteoroids. Some are metallic, magnetic, and very dense. Others are rocky, nonmetallic, and look like ordinary Earth rocks. In some cases, astronomers can determine exactly which asteroid was the source of a meteorite found on Earth by analyzing the make-up of the rock and matching it to known asteroids. Some meteorites have originated from more exotic locations, however. More than 100 meteorites are known to have come from the Moon, and more than 130 Martian meteorites have come from the Red Planet.

## METEORIC BEATDOWN

Meteors pummel Earth. Every day between 10 and 100 tons of material falls into the atmosphere from outer space. The vast majority of it burns up and/or falls into the ocean. Not a single person in the past 100 years has been struck and killed by a meteorite.

## KILLER METEORITES

Meteorites have been known to kill. A meteorite known as Valera came down in Venezuela and reportedly killed a cow in 1972. And on May 1, 1860, a meteorite fell in New Concord, Ohio, and allegedly killed a colt.

## HEAVY HOBA

The largest meteorite on Earth is located in Namibia. It's called the Hoba meteorite and is estimated to weigh about sixty tons. It is the largest single piece of naturally made iron in the world and has never been moved.

# COMETS

### SUN-SURPASSING SIZE

What is the largest thing in the solar system? Usually the answer is the Sun. But occasionally a comet—a dirty snowball of ice and dust—can surpass the Sun in size. The nucleus (the only solid part of the comet) can range from one to twenty miles in diameter, which, astronomically speaking, is very small. But the coma, the envelope of gas surrounding the nucleus, can be 6,000–60,000 miles in diameter. And the tail of a comet can be between 1,000,000 and 350,000,000 miles long. That means some comets, although extremely diffuse, can become even larger than the photosphere of the Sun!

### DON'T LEAVE

A comet like 67P/Churyumov–Gerasimenko has a much smaller mass than a planet or the Moon. The force of gravity is so miniscule on 67P that simply running between 2 and 3 miles per hour would enable you to fly off of it and into space. That means you could take a step and never come down. In fact, you'd be hard-pressed to *not* jump off this comet.

### COMET TAILS

Unlike planets with their relatively circular orbits, comets cut highly elliptical paths around the solar system. Comets spend most of their lives in the dark, cold regions of the outer solar system, but every once in a while they enjoy a fleeting fling with the Sun. As a comet approaches the Sun, the ice heats up and turns into vapor. This vapor forms the coma and leaves a stream of material in its wake—a delicate and beautiful tail.

## A FREQUENT VISITOR

The length of time to complete one orbit around the Sun varies greatly from comet to comet. Short-period comets, like Halley's comet which visits every seventy-six years, complete their orbits around the Sun once every 1–200 Earth years. Astronomers around the world have noted Halley's comet during its periodic passes through the sky since 240 B.C.

## SEE YOU NEVER

Long-period comets are those with orbits lasting longer than 200 years. Comets like Hale–Bopp and Hyakutake were the last two great comets visible in Northern Hemisphere skies when they rounded the Sun in the 1990s. Hale–Bopp will not return to the inner solar system for another 4,200 years, and Hyakutake will not be seen again for another 30,000 years.

## A MILLION-YEAR COMMUTE

The longest-period comets take about 1 million Earth years to circle the Sun one time. The orbits of these comets can take them 1 light year from the Sun, or almost 6 trillion miles, but the gravity of the Sun can still draw some of these comets back for another pass.

## EXILED COMETS

Sometimes comets go around the Sun only once and are thrown out of the solar system. If a comet comes toward the Sun too quickly, it may be ejected into deep space. Or if a comet strays too far from the Sun, it may be snatched away by the gravity of another star.

## INTERSTELLAR COMETS

Astronomers have noted the odd trajectories of certain comets and believe that they may have come from outside the solar system. These rogue comets may have formed around another star system and have traversed the depths of space to find a new home around the Sun.

## DON'T BET ON A COMET

Comets are notoriously fickle objects and defy prediction. Astronomers never know exactly how bright a comet will get. Most brighten as they approach the Sun in a routine way, but some just fizzle out. The greatest astronomical disappointment was Comet Kohoutek. Visible in 1973, the media heralded Kohoutek as the "Comet of the Century." Unfortunately, the comet remained dim and was barely visible to the naked eye. Kohoutek was dubbed Comet Watergate by disillusioned observers.

## NOT SO BRIGHT

In 2013, backyard astronomers were preparing for another potentially spectacular comet to grace the nighttime skies. Comet ISON was big, was going to go really close to the Sun, and could become very, very bright. Unfortunately, ISON underachieved and never got near the magnitude everyone expected. ISON then flew too close to the Sun, melted, and broke into pieces—never to be seen again.

## FINDERS, KEEPERS

Comets are named after their discoverers. Comet hunters are a rare and dedicated breed, but most are amateurs. Japanese amateur astronomer Yuji Hyakutake discovered the comet bearing his name using a pair of large binoculars, and Australian astronomer William Bradfield discovered eighteen comets during more than three decades of searching.

## SHARE AND SHARE ALIKE

Several comets in the 1990s were discovered by two astronomers at the same time, which means the name of the comet had to be shared as well. Comet Hale–Bopp was codiscovered by Alan Hale and Tom Bopp. Likewise, Comet Shoemaker–Levy 9, which crashed into Jupiter in 1994, was the ninth comet discovered by the team made up of Gene and Carolyn Shoemaker and David Levy.

## SCOURING THE SKIES

These days, it is more difficult for amateur astronomers to discover comets. Facilities with larger telescopes such as the Lincoln Near-Earth Asteroid Research (LINEAR) project and the Near-Earth Asteroid Tracking (NEAT) program are finding them first. These projects were designed to monitor asteroids visiting Earth's neighborhood. During their searches, NEAT has found forty-one comets, and LINEAR has tagged 128.

## LIVING ON THE EDGE

In the farthest reaches of the solar system there is a region of space called the Oort cloud. An unknown number of objects, mostly icy, reside in this realm between 2,000 and 100,000 astronomical units from the Sun. Many comets originate from the Oort cloud, and some astronomers believe there may be trillions of other objects out there. The Oort cloud is so far away that astronomers cannot see these objects directly, yet. What lurks in these depths? Astronomers can't wait to find out!

# CHAPTER 8

# SAILING AMONG THE STARS

*The Interesting Lives and
Dramatic Deaths of Stars*

Humanity has always had a preoccupation with the stars. When your ancient ancestors beheld the sky it was full of mystery and wonder. The stars above were otherworldly, heavenly. These earliest sky watchers deified the stars, yet they dared to gaze upon them, to chart them, to classify them, and to use them to tell time. Early humans first painted stars on cave walls. Ancient Egyptians artfully rendered star charts on the insides of coffins and wrote them down on papyruses. Then Medieval scholars wrote books about them, and Renaissance astronomers examined stars with telescopes large and small. Finally, modern astronomers photographed stars and analyzed their light; then they figured out how far away the stars were, what they were made of, how large they were, and if they had planets around them.

No matter how much astronomers discover about the stars, they still exhibit a powerful hold on the imagination. And in this chapter you'll learn how far humanity has come—from wishing upon stars to knowing about their births, lives, and deaths.

# STARS IN HISTORY

## CHART THE STARS

One of the earliest works of human art is also a star chart. A cave painting in Lascaux, France, dating to about 15000 B.C., depicts a bull with long horns and seven spots above its shoulder. Astronomers and archaeologists agree that this is actually a representation of the constellation Taurus and the Seven Sisters star cluster.

## FIRST-MAGNITUDE STARS

Under perfect conditions (clear skies and no light pollution), you can see about 6,000 stars with the naked eye. In cities, however, you may only see the brightest twenty stars in the sky. These incredibly bright stars are called first-magnitude stars.

## NO HUMBLE TELESCOPE

A star's magnitude is a measurement of its brightness. The brightest stars are considered first magnitude. The second brightest are second magnitude, and so on. With each increase in magnitude, the starlight emitted appears 2.5 times dimmer. The human eye can see down to the sixth-magnitude stars. The Hubble telescope, an eight-foot-diameter eye in the sky, can see stars that are thirtieth magnitude, almost 4 billion times fainter than what the eye can see.

## CHANGES IN LATITUDE

The North Star (Polaris) can tell you a lot more than your direction. This star's angle above the horizon in the Northern Hemisphere is also equal to your latitude on Earth. For instance, if you live in New York City, Polaris will be 40 degrees up. In Miami it'll be 25 degrees above the northern horizon. If you were on the North Pole, the North Star would be straight above you. And from the Southern Hemisphere, you cannot see the North Star at all.

## WHO TOLD YOU THAT?

The North Star is not the brightest star in the sky. The North Star ranks about forty-seventh in brightness of all the stars you can see from Earth.

## STAR SCORCHER

The brightest star visible from Earth (other than the Sun) is Sirius. It often twinkles red, white, and blue when it is low in the sky. The ancients called it the "scorcher," and it lives up to its name by blazing almost twice as bright as the second-brightest star, Canopus.

## GOING THE DISTANCE

Astronomers don't use the term *light year* to indicate a really, really long period of time. A light year is a unit of *distance*. It is how far light travels in one year, about 5,878,625,373,183 miles, which is commonly rounded up to 6 trillion miles. Astronomers use light years when measuring the distances to stars, galaxies, and anything beyond the solar system.

## TWO IN ONE, SERIOUSLY

At a distance of 8.8 light years from Earth, Sirius is one of the closest stars to Earth. But it is not alone. Sirius actually is made up of two stars, called Sirius A and Sirius B. Sirius A is the bigger and brighter one and the one you can see with the naked eye. Sirius B is a tiny white dwarf star first seen through a telescope in 1862. Like a planet orbiting a star at a great distance, Sirius B circles around Sirius A once every fifty Earth years.

## DOG DAYS

Sirius is known as the "Dog Star" since it marks the nose of the big dog constellation Canis Major. The phrase "dog days of summer" is also associated with this star. The ancient Greeks noticed that the hottest point of every summer seemed to correspond to the time when Sirius joined the Sun in the daytime sky. This coincidence led to the saying frequently used to describe sweltering July and August afternoons.

## DANCING BEAMS OF LIGHT

Why do stars twinkle? Starlight travels trillions of miles, and when it reaches Earth only a narrow beam of light makes it to your eyes. That beam of light is easily bent when it comes through the atmosphere of Earth, which causes it to dance and twinkle. Stars twinkle more when they appear low in the sky. When this happens, the starlight must go through more of Earth's atmosphere to reach you. More atmosphere equals more twinkle.

## A KNOWN TWINKLER

The most notorious twinkling star is called Capella. When Capella is low in the sky, its light appears to change color, flicker, and dance dramatically. The light you see doesn't come from just one star but from four stars orbiting each other—two yellow suns and two little red ones.

## EGYPTIAN PIVOT POINT

Five thousand years ago, Polaris was not the North Star. Due to the slow wobble of Earth, called precession, a star in the constellation Draco, named Thuban, held that honor. Thuban is not as bright as Polaris, but the ancient Egyptians recognized it for being the pivot point of the sky.

## KOCHAB, GUIDE THE WAY

Around 1900 B.C., the North Pole of Earth pointed more directly at the star Kochab in the constellation Ursa Minor. Kochab was the polestar for hundreds of years and helped guide travelers.

## ANCIENT PERSIAN ROYALTY

Ancient Persian astronomers broke the sky into four sections, each ruled by a royal star. These royal stars were Regulus in the constellation Leo, Antares in Scorpius, Fomalhaut in Piscis Australis, and Aldebaran in Taurus.

## EPSILON EXCEPTION

Almost every star you see at night is much larger than the Sun. There is one exception. A star called Epsilon Eridani is the only star that you can find with the naked eye that is smaller than the Sun. Despite its small stature, Epsilon Eridani is visible since it is one of the closer stars at only 10.5 light years away.

## THE HEART OF THE SCORPION

The bright blue star called Vega is 2.5 times wider and 40 times brighter than the Sun. Another notable star, Arcturus, has a diameter 25 times that of the Sun and shines 170 times brighter. But one of the true behemoths in our part of the Milky Way is Antares, the red supergiant star that marks the heart of the constellation Scorpius. Antares is 883 times wider and more than 57,000 times brighter than the Sun.

## BETTER THAN AVERAGE

The Sun is way above average in size and brightness and more massive than 80 percent of the known stars. If you include all the stars you can see in telescopes, most of these stars are surprisingly small. A category of stars called red dwarfs are all over the place; however, you can't see any red dwarfs with the naked eye.

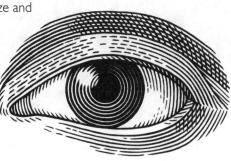

# STAR QUALITIES

## RAINBOW BRIGHT

When you think of stars, you may immediately think they are all white. Not so. Stars come in all colors of the rainbow, from ruby red to tangerine, from blonde to deep blue. You can tell a lot about a star by its color.

## RED HOT, BLUE HOTTER

The color of a star is determined by the star's temperature. If you think of the spectrum of colors from blue to red, the hotter stars are on the blue side and the colder stars are on the red side. The color spectrum of stars from coldest to hottest is red, orange, yellow, white, and blue.

## DEEP BLUE BELLATRIX

Astronomers assign most stars to one of seven categories designated by a letter. These seemingly random designations are O, B, A, F, G, K, and M. O stars, like the blue Bellatrix, which has a surface temperature of almost 40,000°F, are incredibly hot stars. A red star like Betelgeuse is a cool 5,400°F and falls under the category of an M star.

## OH, BE A FINE GUY OR GIRL

Astronomers have come up with two catchy phrases to help remember the seven stellar categories (OBAFGKM). Depending on your preference you can say, "Oh Be A Fine Girl, Kiss Me" or the alternative, "Oh Be A Fine Guy, Kiss Me."

## A G-CLASS STAR

Astronomers classify stars by temperature and size. The Sun ranks as a G2V star. The "G2" part is the Sun's spectral class identifying it as yellow-white color with a surface temperature of 10,000°F. The "V" indicates that the Sun is a main sequence star. When astronomers look for planets around other stars, they look at other G2V stars very carefully.

## THROUGH THE LOOKING GLASS

It's never boring to look through a telescope at an individual star. With a little extra magnification, the colors become much more distinct. The bright stars Vega and Rigel are brilliant blue in a telescope. And optical aids can bring out the red in the reddest stars like Mu Cephei (also known as Herschel's Garnet Star) in the constellation Cepheus and La Superba in the constellation Canes Venatici.

## OUT IN THE COLD

The coldest star that has been measured is named WISE 0855−0714. It's a small brown dwarf star that's 7.2 light years from Earth (one of Earth's closest stars). Its surface temperature is estimated to be between −55°F and 8°F. How can a star be colder than ice? Is a brown dwarf just a really large planet? Welcome to the weird, not-fully-understood world of brown dwarf stars.

## TOO HOT TO HANDLE

The hottest known star is a small dynamo called WR 102. Even though WR 102 is no more than 40 percent the diameter of the Sun, it is a whopping 210,000°F! This star can be found in the constellation Cygnus but is so distant and faint to Earth that you'd need a very large telescope just to find it.

## IT'S NOT EASY BEING GREEN

Some say that there aren't any green-colored stars, but this is open to individual interpretation. Fomalhaut, the brightest star in the constellation Piscis Australis, sometimes appears to twinkle green. Find Fomalhaut in your fall evening sky and see what you think.

## THE EYE OF SAURON STAR

If you look at certain pictures of the star Fomalhaut, you'll notice a resemblance to the Eye of Sauron from the Lord of the Rings trilogy. This star is so bright that astronomers mask out its glare in order to see what's around it. When they did this, they found rings and rings of dust circling around Fomalhaut. Embedded in that dust was a planet, looking like a slow-moving lump around the Eye of Sauron. This was the first planet to be seen visually orbiting another star.

## HOW FAR IS THAT STAR?

Astronomers use parallax, the apparent movement of a star relative to background stars, to measure distances to nearby stars. They look at a star once and then measure its position again six months later. The bigger the movement, the closer the star. In 1838, German astronomer Friedrich Bessel became the first person to successfully measure the distance to a star (he calculated the distance to 61 Cygni, which turned out to be 11.36 light years away) using stellar parallax.

## DISTANT DENEB

The star Deneb may be the farthest star you can see with the naked eye. It is an estimated 3,000 light years away (about 17,000,000,000,000,000 miles). Because it is one of the brightest stars you can see on a summer and fall evening, it must be humongous. Deneb could be 50,000–200,000 times brighter than the Sun.

## A SQUISHED STAR

Altair, the brightest star in the constellation Aquila, spins crazy fast. While the Sun rotates once every twenty five to twenty-six days, Altair completes one turn every nine hours. This rapid rotation has flattened it at the poles and sent more mass to its equator. It looks more like a piece of M&M's candy than a sphere.

## MORE THAN MEETS THE EYE

Most of the stars in the sky are not just one star. They consist of two, three, or more stars orbiting each other. These double-, triple-, and multiple-star systems are the norm, while our single, lone Sun is in the minority in its section of the Milky Way.

## COMPLEMENTARY COLORED STARS

With even a simple telescope you can examine a beautifully colored double star (a system of two suns orbiting each other) of contrasting colors named Albireo. The brighter of the two stars is orange with a dimmer blue star orbiting it. Albireo, which is visible to the naked eye, marks the head of Cygnus the Swan. Cygnus, also known as the Northern Cross for its resemblance to a crucifix, is easy to identify high in the sky during the summer and fall evenings.

## TRILLIONS TO TRAVEL

At about 93 million miles away, the Sun is Earth's closest star. The second closest is Proxima Centauri, which is about 25 trillion miles from Earth.

## TRIPLE THREAT

You cannot see the second-closest star, Proxima Centauri, with the naked eye, but you can see the system in which it belongs. Proxima Centauri is one component of a three-star system called Alpha Centauri. Two yellow suns circle each other while Proxima, a tiny red dwarf star, circles the other two. If you lived on a planet around one of these stars, you would see three suns in the sky and could watch one sun set in the west while another rises in the east.

## GETTING THERE IS MORE THAN HALF THE FUN

It took the Apollo astronauts about three days to fly to the Moon. It would take you about seven months to fly to Mars. It took the New Horizons, the fastest spacecraft ever launched, nine and a half years to fly past Pluto. But if you wanted to fly to the nearest star outside the solar system, Proxima Centauri, with New Horizons it would take you more than 74,000 years to get there.

## SEXTUPLE AND SEPTUPLE STARS

The bright star Castor, in the constellation Gemini, consists of six stars that orbit each other. However, two other stars may have Castor beat. Astronomers believe that the stars named Nu Scorpii and AR Cassiopeiae may each be septuple stars—with seven stars circling a common center. If you lived on a planet around one of these stars, would it ever get dark?

## WIDE-OPEN SPACES

Stars are also found in clumps in some parts of the sky. Astronomers call the smaller clumps of stars "open clusters" since they can see more open space between the stars. There can be hundreds of stars in an open cluster.

## THE BACK OF THE BULL

There are two open clusters visible to the naked eye in the fall and winter sky: the Hyades and the Pleiades. The Hyades form the face of the constellation Taurus the Bull, while the Pleiades (AKA the Seven Sisters) ride on the bull's back.

## STAR GLOBS

Larger clumps of stars are called globular clusters and they look like globes of light in a small telescope. But in a larger scope, you can see the thousands of individual stars like a swarm of fireflies filling your entire view. When you observe a globular cluster you are seeing the light of hundreds of thousands of stars.

## IT'S COMPLICATED

Within the large clumps of stars called globular clusters the stars are more densely packed than in most sections of the Milky Way. This creates complicated gravitational interactions between the stars and may make an extremely difficult environment in which planets could form. But even within a globular cluster, space is mostly empty and stars do not come close enough to collide with one another.

## OLDER THAN DIRT

Globular clusters are some of the largest structures observed within the Milky Way as well as some of the oldest. Within the Milky Way, some globular clusters contain stars that are more than 11 billion years old. They may have been the first stars to form in the earliest stages of the Milky Way.

## ONE GIGANTIC STAR SYSTEM

Astronomers call billions to trillions of stars orbiting around a common center a galaxy. The Sun resides with a swarm of other stars in the Milky Way galaxy that has between 200 billion and 1 trillion stars in it.

## LOOKING AT THE MILKY WAY

How do astronomers know what the Milky Way looks like? A spacecraft has never been sent outside the galaxy in order to take a picture of it from the outside. So astronomers must map the stars and star clusters around Earth and plot them all in three dimensions. With enough data they can estimate what the Milky Way would look like from a galaxy far, far away.

## MILKY MYTHOLOGY

You may not be able to see what the Milky Way looks like from outside the galaxy, but it appears in the nighttime sky like a trail of spilled milk. The origin story for the name *Milky Way* comes from the ancient Greeks. In Greek mythology the goddess Hera was breastfeeding baby Hercules when she learned that she was not his mother. Hercules was the son of her husband Zeus and the mortal woman, Alcmene. As Hera pushed Hercules away from her, a spurt of milk flew out and made the Milky Way. Even in modern Greek, the word *gala* means "milk."

## THAT'S A LOT OF STARS . . .

How many stars are there in the universe? Astronomers are not sure—they're still counting them. But some estimates show that there could be 300,000,000,000,000,000,000,000 stars in the universe.

# PICTURE THIS!

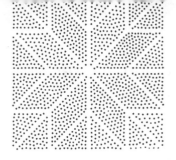

## PATCHWORK OF SPACE

A constellation is a group of stars that, with mild to considerable imagination, can form a picture in your mind. There are eighty-eight officially recognized constellations in modern astronomy. Most of these, like Orion, Leo, and Sagittarius, were named by the ancient Greeks thousands of years ago. The International Astronomical Union (IAU), the authoritative body of astronomers, established the eighty-eight official constellations in the 1920s. The IAU also set their boundaries, which look like a patchwork quilt of rectangles that cover every square inch of the sky.

## SO MUCH FOR THE POPULARITY CONTEST

Some legendary constellations proved unpopular and did not make it in the official eighty-eight recognized by the IAU. These former constellations include Noctua the owl, Vespa the wasp, and Cerberus the three-headed dog of the underworld. Sometimes you can find drawings of their starry likenesses on old star maps.

## UNLICENSED STAR SALESPEOPLE

Although you can "buy" a star from a variety of private companies, you cannot purchase a star from the IAU. They are the only official naming body in the world that is recognized by astronomers and the scientific community. So if the IAU doesn't name it for you, it's not legit.

## THE EYE OF THE BEHOLDER

The most recognizable star pattern in the Northern Hemisphere is the Big Dipper. To American sky watchers, seven semibright stars seem to form the shape of a spoon. Other cultures saw different pictures in these stars, such as a fishhook, a plow, the thigh of an ox, a stretcher, a cart, a coffin, and a drinking gourd.

## CELESTIAL TIMEPIECE

If you live in the Northern Hemisphere, you can see the Big Dipper in the sky almost all year round. Every night it circles so close to the North Star that it sometimes appears to be standing on its handle, and other times it seems to be standing on its spoon. Sometimes it looks to be holding water and other times that it is dumping out its contents. Ancient astronomers could use the position of the Big Dipper to tell time. It makes a great night clock.

## UNOFFICIAL GUIDE TO THE SKY

The Big Dipper is not a constellation. It is only one section of a larger star picture called Ursa Major, the Big Bear. The Big Dipper is an asterism, an unofficial star pattern. It makes up the rear end and tail of a mother bear while other stars mark her feet and head to complete the constellation.

## BIG-CITY CONSTELLATIONS

Asterisms, or unofficial star patterns, are also referred to as urban constellations. The Big Dipper along with other asterisms like the Summer Triangle and Orion's Belt are easier sights to see in light-polluted skies than the entire constellations around them.

## IT'S THE BIG DIPPER, NO DOUBT

You see a dipper in the sky, but is it the Big Dipper or Little Dipper? If you see only one dipper in the stars, it is the Big Dipper. The Little Dipper is almost impossible to see in its entirety from urban or suburban locations because its stars are so faint. The Big Dipper is so much more distinct than its smaller, dimmer counterpart.

## DIPPER DIRECTIONS

The Little Dipper is the asterism inside the constellation Ursa Minor, the Little Bear. The North Star marks the end of the stretched-out tail of Ursa Minor and can be found with help from the Big Dipper. Connect the dots of the two stars on the end of the Big Dipper's spoon. Continue that line of sight away from the spoon and you will run into Polaris, the North Star.

## ALONG THE 40TH PARALLEL

Your perspective on the stars doesn't change when you travel east and west. That means that everyone who lives within a few degrees of 40 degrees north latitude sees the same stars and constellations. Viewers from New York to San Francisco, Japan to Central Asia, and even southern Europe and North Africa have pretty much the same view of the cosmos.

## ONE DEGREE OF SEPARATION

Your perspective on the stars changes if you travel north or south. If you live in the Northern Hemisphere, when you head south you're travelling down the curve of Earth. You get to see stars that were previously blocked by your southern horizon. For every 1 degree of latitude you travel south, you'll get to see 1 extra degree of the southern sky.

## NORTH OR SOUTH?

The ancient Greeks, Romans, Mesopotamians, Indians, and Chinese could not see all the stars in the sky. Since they lived in the middle latitudes of the Northern Hemisphere, there was a good portion of the sky that never rose above their southern horizon. Cultures in Central and South America, Africa, and Oceania named these constellations and had their own mythologies behind them. Only people who live exactly on the equator can see all the constellations in the northern and southern sky.

## NOT-SO-ANCIENT NAMES

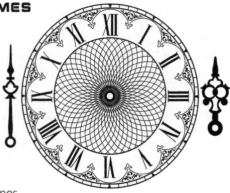

Western Europeans first saw the stars of the Southern Hemisphere during their voyages of exploration in the 1500s. They added these "new" stars to their charts, and some constellations visible in the Southern Hemisphere now bear less ancient names, like Telescopium, Microscopium, and Horologium (a clock). French astronomer Nicolas-Louis de Lacaille named these constellations in the eighteenth century.

## TIME TO HEAD FOR HOME

Sailors from western Europe often grew very nervous traveling south along the coast of Africa. The farther south they went, the lower Polaris got in the sky. When sailors approached the equator, the North Star was barely visible. Many sailors turned back and headed for home so that they would not lose sight of the North Star and their bearings.

## STAR-FREE SOUTH

There is no "South Star." While there is a polestar in the Northern Hemisphere (the North Star), there is no equivalent in the Southern Hemisphere. As the night goes on, the stars appear to circle an area of space that is seemingly free of bright stars.

## CELESTIAL GPS

To help tell directions in the Southern Hemisphere, travelers can use the Southern Cross. The longer part of the cross (the two stars furthest apart) will point you toward the south celestial pole (where a "South Star" would be). Travelers in the Sahara desert also used the second-brightest star in the sky, named Canopus, to guide them across North Africa.

## WHAT'S YOUR SIGN?

The Sun appears to pass in front of several famous constellations: Aries, Taurus, Gemini, Cancer, Leo, Virgo, Libra, Scorpius, Sagittarius, Capricornus, Aquarius, and Pisces. This is the zodiac, and astronomers use these twelve constellations as markers of space and time. The path that the Sun appears to take is called the ecliptic—since those are the only places where eclipses can occur.

## STAR SERPENT

The largest constellation in the entire sky—Northern **or** Southern Hemisphere—is Hydra, the many-headed serpent that battled Hercules in the Greek myths. From heads to tail, Hydra spans more than half of the sky and can be best seen on April evenings.

## SOUTHERN EXPOSURE

The smallest constellation in the sky is Crux, the Southern Cross. You can't see this constellation from locations north of the 26th parallel, but these four bright stars make this tiny constellation a highlight of the Southern Hemisphere sky. The Maori people of New Zealand called these stars "the anchor," while the ancient Inca of Peru referred to these stars as "a stair."

## THIS WAY TO THE CENTER OF THE GALAXY

When you find Sagittarius in the sky, take special note. Peer deeply past the stars and into the further blackness of that space. In that direction, 25,000 light years away, is the center of our galaxy—the center of the Milky Way.

## IT'S NOT ALL GREEK TO ME

Most constellation names come from ancient Greece, but most star names, like Betelgeuse, Rigel, Algol, and Zubenelgenubi, were named by ancient Arabic astronomers. During the Dark Ages in Europe, scholars in the Muslim world translated, updated, and improved upon the ancient Greek astronomy texts and left their marks on the stars' names.

## A SNAPPY DRESSER

Who's the star guy wearing the snazzy belt of three stars in a row? The most famous and easy-to-recognize constellation: Orion the Hunter. Orion is the constellation that conjures the deepest imagination and wonder with just one glance. Something about the placement of the stars ties the entire picture together. Almost every culture in the ancient world associated these stars with a hunter, giant, or all-around he-man.

## THE CENTRAL ONE

Arabic astronomers took special interest in the constellation picturing Orion the Hunter. They referred to him as "the Central One" for his prominent place in the North African sky. Astronomers today still refer to many of Orion's bright stars by their Arabic names.

## THE SUN BELT

The recognizable three stars in the belt of the constellation Orion bear Arabic names. They are called Alnitak, Alnilam, and Mintaka. Although these names sound dynamic, respectively they mean "String of Pearls," "the Girdle," and simply, "the Belt."

## ORION'S FEMININE SIDE

The star that marks Orion's left shoulder is called Bellatrix. *Bellatrix* means "beautiful warrior woman" or "Amazon star." How this star, attached to the most macho of all constellations, got this name is still a mystery.

## DON'T SAY IT THREE TIMES

The most famous star in Orion is Betelgeuse. No one knows exactly how to pronounce many ancient star names since they were changed so many times over the years. So, say Betelgeuse however you want. If you want to sound snooty, say, "bet-el-GEEZE," or have fun with it and call it, "BEETLE-juice."

## BODY OF STARS

The intended meaning of Orion's famous star Betelgeuse is "armpit of the central one," and it is found, appropriately, under his upraised right arm. Orion's other bright star, Rigel, means "left foot" and is placed as such.

## SUPER-COOL BETELGEUSE

Orion's star Betelgeuse is a red supergiant star about 640 light years from Earth. It shines with a ruddy orange color and therefore is a cooler star than almost any other noticeable star in Orion.

## A HOT TIME IN THE OLD STAR TONIGHT

Betelgeuse is humongous. If Betelgeuse was placed where the Sun is, its sweltering atmosphere would reach the orbit of Jupiter, which means Mercury, Venus, Earth, and Mars would all orbit inside Betelgeuse. Life on Earth would be pretty tough inside a 5,400°F cauldron of gases.

## BETELSPOTS

Orion's star Betelgeuse is so big it is one of the few stars that you can see well from Earth. In telescopes, almost every star just looks like a pinpoint of light. But with Betelgeuse you can see a weird-shaped disc and features on its surface. It even has sunspots, or should you call them "Betelspots"?

## END OF DAYS

Betelgeuse is a variable star. It dramatically fluctuates in brightness and size. The reason for all these big changes is that Betelgeuse is nearing the end of its life. The fluctuations are caused by the push and pull between the forces of gravity (wanting to hold the star together) and internal pressure (trying to blow the star apart). Therefore, Betelgeuse will go supernova soon when it will explode in a blast so violent that it will be visible in our daytime sky. Now when astronomers say "soon," they could mean tomorrow or within the next 10,000 years.

# WEIRD STARS

## AND THE WINNER IS . . .

What is the largest star? This is not an easy question to answer. Because stars are so far away and tough to resolve perfectly, astronomers are still debating this question. A star named VY Canis Majoris previously held the title of largest star in the galaxy, but now several other contenders are vying for the crown. Leading the way is UY Scuti, a star 1,700 times the diameter and 5 billion times the volume of the Sun.

## PUPPIES FOR THE DOG STARS

Many stars have smaller companion stars that circle around them. These tiny but brilliant stars are called white dwarfs, and Sirius, the Dog Star, as well as Procyon, the Little Dog Star, each have one. White dwarfs are super-dense objects. They pack the mass of the Sun into something the size of Earth.

## ETERNAL LIGHT

Astronomers think that the vast majority of stars (including the Sun) will eventually evolve into white dwarfs. From here, they will live a very, very long time—possibly 100 nonillion years. That's 100 followed by thirty zeros.

## DUSTY DISK

At only about 12 million years old, Beta Pictoris is an extremely young star. When astronomers detected a huge disc of dust surrounding it, they were looking at an early version of the solar system. The planets and asteroids orbiting the Sun formed from such a disc billions of years ago. Beta Pictoris has at least one planet circling it, as well as many hundreds of comets.

## SOME KIND OF WONDERFUL

Mira is a giant red star in the constellation Cetus. Normally Mira isn't even visible to the naked eye. But for a few days every year Mira shines brighter and even acquired the nickname "the Wonderful" by those who first witnessed her flare-up. The difference between normal Mira and bright Mira is startling. The greatest flare-up of Mira brought her to 1,500 times her normal brightness.

## SHEDDING STAR PARTS

The giant red star Mira is actually a binary star—two stars with a smaller one orbiting the larger one. As the stars age, the smaller star is actually stealing mass from the larger star. NASA's GALEX satellite saw that Mira is casting aside mass as it flies through space at over 80 miles per second. It's leaving something like a comet's tail as it goes that is now more than 13 light years long.

## FREQUENTING THE SEQUENCE

Most of the stars in our part of the galaxy, including the Sun, fall into the category of main sequence stars. When two astronomers, Ejnar Hertzsprung and Henry Norris Russell, plotted the luminosity and tem-perature of stars they found a strong correlation. About 90 percent of the stars fall on this main sequence and share a similar luminosity-to-temperature ratio.

## IT'S ALL ABOUT MASS

The lifespan of a star is determined by its mass: the more massive the star, the shorter the lifetime. Big stars will live millions of years, medium-sized stars will last billions of years, and small stars can shine on for trillions of years.

## STARRY DEMISE

The mass of a star will determine the magnitude of its demise. Small stars will just cool and fade away. Medium-sized stars will expand until the stars can't hold themselves together. The outer layers of the star just blow away and don't come back. These gases form what's called a planetary nebula (since it looked like a round planet through old telescopes). And massive stars die spectacularly. They explode into supernovas and/or implode into black holes. For a brief time, a supernova releases so much energy that it can outshine an entire galaxy of billions of stars.

## TWO SUNS IN THE DAYTIME SKY

On April 30, 1006, a dazzlingly brilliant, new star appeared in the sky. It outshone every other star, was more than ten times brighter than Venus, and could easily be seen in the daytime. This was the supernova of 1006, the brightest star death ever recorded.

## OVERDUE SUPERNOVA

The last super-bright supernova (the death of a huge star) occurred in 1604. This explosion was so bright that it turned a faint star into something that could be seen during the daytime. Imagine seeing the Sun in the sky accompanied by another star! When will the next supernova occur? It is anyone's guess.

## GHOULISH ALGOL

Ancient stargazers from around the world noticed something strange about the star Algol. Sometimes it was bright and other times it was dim. Whatever was happening to Algol was interpreted as a bad omen. So the star was called, by various cultures, "the Ghoul," "Medusa's Head," "Satan's Head," "Demon Star," and "Piled-Up Corpses." In reality, Algol is two stars that revolve around each other. When one star eclipses the other, it blocks out a lot of light pointed toward Earth. You can see this dip of starlight for ten hours every 2.86 days (the time it takes one star to orbit the other).

## DONUT STAR SYSTEM

The light from the star Epsilon Aurigae dips by almost 50 percent every twenty-seven years. Why? Astronomers think that a dark mass of material orbits this star and regularly eclipses its light. Or there is a dusty disc that encapsulates the star like a dark donut. The next dimming of Epsilon Aurigae is expected around 2037.

## STAR LIGHT, STAR BRIGHTER

P Cygni is a luminous blue variable star and one of the brightest stars in the Milky Way. No one noted it until 1600 when it flared up to naked-eye brightness. P Cygni faded away and then brightened again several times until 1715. It has remained a dim star in the sky since then, but if the conditions are right, you can see it across 5,500 light years of space.

## FAIR ILLUMINATION

The star Arcturus triggered the start of the 1933 World's Fair in Chicago. When this star's light fell on a sensor, it activated a switch that illuminated the fairgrounds on opening night.

## KEEP IT QUICK

Barnard's Star appears to change its position within the constellation Ophiuchus faster than any other star in the sky. This effect is called proper motion, and from Earth's perspective, Barnard's Star shifts 1 degree in space about every 350 years.

## THE METHUSELAH STAR

HD 140283 was nicknamed "the Methuselah star" after astronomers first determined it to be more than 14 billion years old. This could not be correct since the universe has only been around for 13.8 billion years. How could a star be older than the universe? Further measurements brought HD 140283's age back into the realm of possibility—somewhere in the 13-billion-years-old range. As of 2016, the oldest star is thought to be HE 1523-0901, a red giant almost 13.2 billion years old.

## DIZZY PULSAR

A pulsar is a small, rapidly spinning star that emits strong bursts of radiation into space. PSR J1748-2446ad wins the prize for fastest-spinning object in the galaxy. This ten-mile-wide pulsar spins over 700 times per second.

## THE ZOMBIE STAR

A Thorne–Żytkow object is a star within a star. When a small neutron star collides with a humongous red star, it can be absorbed and still live on. Astronomers believe that they found one in 2014 in the form of HV 2112, a massive star in the nearby galaxy called the Small Magellanic Cloud.

# CHAPTER 9

# DEEP SPACE

*Highlights and Strange Sights in the Milky Way and Beyond*

In 1758, a twenty-eight-year-old French astronomer named Charles Messier beheld a marvelous sight. Halley's comet had returned as predicted and proved that these fuzzy-tailed visitors orbited the Sun like the planets. Later nicknamed the "Ferret of Comets," Messier discovered thirteen comets and codiscovered six others. However, Charles Messier is known today for his list of deep-space objects, not for his comets. In his search for these extremely faint objects, Messier came across areas of the sky that looked like comets but did not move with respect to the stars. Messier called these comet masqueraders "embarrassing objects."

So Messier, not wanting to be embarrassed anymore, compiled a list of comet-like objects on star charts as a sea captain might mark reefs on an ocean map. Eventually, he included 103 objects in his catalog, and seven more were added after his death.

These 110 Messier objects represent some of the most fascinating phenomena in and out of the Milky Way. They include anything outside of the solar system and objects more complex than individual stars. They are also called deep-sky objects and include more massive structures such as star clusters, nebulas, and galaxies, and you can find a lot of these amazing celestial gems for yourself with only minimal equipment. In this chapter you'll investigate some of the most interesting Messier objects and then go even farther and follow other explorers of the universe and delve into the weird, wild world of black holes. Your exploration of deep space begins now!

# CATALOGING DEEP-SPACE OBJECTS

## CODE NAMES

Astronomers designate the brighter deep-space objects with the prefixes M (Messier), NGC (New General Catalogue), and IC (Index Catalogue). M objects were cataloged by Charles Messier in the eighteenth century and include 110 of the brightest and most impressive-looking objects in a telescope. There are more than 7,000 NGC objects and 5,000 IC objects, which were all observed and mapped by the Danish astronomer John Louis Emil Dreyer in the late nineteenth and early twentieth century.

## MESSIER'S MIXED BAG

The Messier Catalog includes forty galaxies, twenty-nine globular star clusters, twenty-six open star clusters, eleven nebulas, one starry patch of the Milky Way, one double star, one star pattern, and one supernova remnant. Excluding galaxies, Messier noted seventy fascinating and scientifically important objects that lie within the Milky Way.

## WITHIN YOUR SIGHTS

About fifty Messier objects can be found with a good pair of binoculars, and the rest are within reach of telescopes that have a lens or mirror eight inches in diameter or greater. But despite the pretty, colorful pictures you see of deep-space objects, they don't look very impressive in ordinary telescopes. They look like faint, ghostly gray shadows, fuzzy blobs, or sometimes like a really faint comet.

## MESSIER'S FIRST

Charles Messier described the origins of his catalog this way: "What caused me to undertake the catalog was the nebula I discovered above the southern horn of Taurus on September 12, 1758. This nebula had such a resemblance to a comet in its form and brightness that I endeavored to find others, so that astronomers would no more confuse these same nebulae with comets just beginning to appear." This object became known as M1.

## SUCH A CRAB

Astronomers know that M1, or the Crab Nebula, is the remnant of a supernova explosion that lit up the sky in the year 1054. This supernova was so bright that it could be seen in the daytime. M1 through a modern telescope looks like the same fuzzy blob Messier observed in 1758.

## NO TELESCOPE NEEDED

Some Messier objects can be seen with the naked eye. You can see the Orion Nebula (M42), the Beehive Cluster (M44), and the Pleiades (M45) without a telescope.

## THE BABY MAKER

M42, the Orion Nebula, is a cloud of gas and dust 24 light years wide that is creating new stars. It is a huge star factory. Gravity is amassing gases from this section of the galaxy into huge clouds. When enough material comes together, the temperature inside the cloud starts to slowly rise. When individual pockets of gas reach a certain temperature and pressure, nuclear fusion begins and a star is born. Can astronomers predict when a new star will be born from the Orion Nebula? No. Star formation is a slow process that takes eons, and astronomers have to watch and wait.

## TOURING THE STAR FACTORY

The Orion Nebula has enough material in it to create about 2,000 Suns. The Hubble Space Telescope has observed newborn stars within M42, some with shadowy discs of material around them. These discs may be the precursors to families of planets that are circling on a flat plane like the planets in the solar system.

## BELOW THE BELT

Even though it is about 1,500 light years away, M42 is the closest large, star-forming region to Earth. It is so large that you can see it with the naked eye as a fuzzy star-like object in the constellation Orion. M42 is found below Orion's Belt of three stars, and is the center of three other stars that form Orion's Sword.

## BORN FROM THE CLOUD

Through a backyard telescope you can resolve individual stars in M42 that are surrounded by a gray cloud of material. Four blue stars, called the Trapezium, lie in the center of the Orion Nebula and light up the surrounding cloud. These stars even illuminate the edges of M42, which look like tendrils of gas reaching out into space. The Trapezium stars were made only thousands of years ago—a super-short time astronomically speaking.

## A BELT LOOP

Circling around the Orion Nebula is a huge, red, ghostly curlicue of gas called Barnard's Loop. This loop stretches hundreds of light years across space and was most likely the result of a supernova explosion 2 million years ago.

## DOWN THE DUSTY LANE

M43 is a section of M42 separated by a dark dust lane of gases. Through a telescope, it looks like an additional globe of light near the main body of the Orion Nebula. The light you see is ionized hydrogen and an indicator of recent star formation.

## THE NORTHERN AND SOUTHERN ASSES

The ancient Greeks and Romans called M44 (the Beehive Cluster) *Prae-sepe*, meaning "the manger." It can be found between two faint stars in the constellation Cancer: Asellus Borealis (northern ass/donkey) and Asellus Australis (southern ass/donkey). Roman legends describe these stars as two donkeys eating from the manger that is M44.

## BEHOLD THE BEEHIVE

M44 is sometimes called the Beehive Cluster since some people say that the light of its stars resemble a swarm of bees circling a hive. It is an open cluster consisting of 1,000 stars with a common center of mass.

## A PLANETARY DEBATE

Astronomers discovered two planets in M44. Each orbits a different star within the massive cluster. This answered a debate among astronomers as to whether planets could form within the complex gravitational influences of star clusters. They can!

## THE SEVEN SISTERS

M45 is often called the Pleiades, or the Seven Sisters. They are an open cluster of stars about 440 light years from Earth. The Pleiades cover an area about as large as a full moon. At first glance they look like a cloud, but when you peer deeper you can see individual stars in the cluster. If conditions are right, you can see seven stars (the seven sisters) with the naked eye. Without binoculars on most nights in more urban locations you can only see four, five, or six of the seven brightest stars in the cluster, shaped like a miniature version of the Big Dipper—a Dinky Dipper, if you will.

## KNOW YOUR SISTERS

The individual stars in the Pleiades are named for figures in ancient Greek mythology. The Seven Sisters are named Alcyone, Maia, Taygeta, Sterope, Merope, Celaeno, and Electra. These named stars form the cup shape of the cluster and are joined by their parents, Atlas and Pleione, the stars that add a stubby handle to the sisters' cup.

## WHAT MAKES A SUBARU?

In Japan, the Pleiades are called Subaru. Check out the logo for the car company Subaru and you'll find the pattern of these stars accurately depicted.

## STAR MASS

Through a pair of binoculars you can see about fifty stars in addition to the Seven Sisters within the Pleiades. Astronomers have found hundreds more through their telescopes and estimate the mass of the entire cluster to be 800 times that of our Sun.

## BURNING THE CELESTIAL CANDLE

The stars in the Pleiades formed within the last 100 million years from a huge nebula. Most of the stars are blue in color and thus much hotter than the Sun. They will also have a much shorter lifespan than the Sun and blaze out in hundreds of millions of years. Astronomers often characterize them as young, hot stars burning the candle at both ends.

## ROUND IT UP!

Why did Messier include M42, M43, M44, and M45 in his catalog even though everyone could see that they weren't comets? When he made his first publication of noncomets, he had only forty-one objects to showcase (the remaining sixty-five came later). Messier wanted a rounder number and added the four unique interstellar features in order to reach forty-five objects.

## MORE WHERE THAT CAME FROM

Another nebula in the constellation Orion is called M78. While not as big and bright as the Orion Nebula, M78, even at 1,600 light years away, is still observable in a good telescope. Look for it above Orion's left belt star, Alnitak. Two star-forming regions light up the cloud with a dark lane of dust dividing them.

## RING AROUND THE WHITE DWARF STAR

M57 is also known as the Ring Nebula. Through a telescope, it looks like a little smoke ring set against the blackness of space. However, M57 is another example of a planetary nebula formed when a white dwarf star was created. Its gases span about 2.6 light years of space, and it resides about 2,300 light years from Earth.

## OWL IN THE LEVIATHAN

M97 is called the Owl Nebula and was nicknamed by English astronomer William Parsons. In 1848, after Parsons saw M97 through the enormous reflecting telescope he called "The Leviathan of Parsonstown," he drew a picture of the nebula. It looked so much like the head of an owl that the name has stuck ever since.

## MIGHTY HERCULES

One of the largest globular clusters visible in the Northern Hemisphere is called M13, or the Hercules Globular Cluster. It has about 300,000 stars and stretches across almost 170 light years of the galaxy. But M13 is so far away (about 2,500 light years) that you can just barely see it with the naked eye under a dark sky.

## WHAT CAME FIRST?

Globular clusters are some of the oldest structures in the Milky Way. M13, for instance, is estimated to be more than 11 billion years old. Do globular clusters predate the Milky Way? Were they around before our galaxy formed? Some astronomers believe so.

## PTOLEMY AND THE BUTTERFLY

M6 and M7 are Messier objects that can be seen with the naked eye under a dark sky. They are both open clusters above the tail stars of Scorpius the Scorpion. M6 is called the Butterfly Cluster since it looks a little like a butterfly of stars when viewed through a pair of binoculars or small telescope. M7 is nicknamed the Ptolemy Cluster since the ancient Greek astronomer Ptolemy noted it in A.D. 130.

## THE WILD DUCKS

M11 is one of the more densely packed open clusters and consists of about 2,900 stars circling a common mass. It is also called the Wild Duck Cluster since the stars, through a telescope and depending on your imagination, form the shape of one to several ducks flying through space.

## NEBULOUS NEBULAS

The astronomical term *nebula* is actually very nebulous. In Charles Messier's day, a nebula was any object that looked cloudy in a telescope. Today astronomers still use this term to describe a variety of objects, including star-forming regions like the Orion Nebula as well as areas where stars are dying like the Ring Nebula.

## TRIFECTA IN THE TRIFID

The Trifid Nebula (M20), one of the best deep-sky objects to find in a telescope, has a unique combination of features including an emission nebula (a glowing pink bulb), reflection nebula (a glowing blue bulb), and dark nebula. The emission nebula is where gases from the open star cluster are ionizing energetically. Within this pink section are dark streaks that give M20 a rose-like appearance. These streaks, made of molecular clouds of gas and dust that give the appearance of empty space, are called dark nebulas or absorption nebulas. The reflection nebula is where distant interstellar clouds of dust are reflecting the light of the stars.

## THE GRAY LAGOON

The Lagoon Nebula (M8) is a bright star-forming region. Although not nearly as bright as the Orion Nebula, you can see M8 with the naked eye under an extremely dark sky. It will look like a faint gray cloud in the constellation Sagittarius.

## FROM THE EAGLE'S MOUTH

M16 is also known as the Eagle Nebula since its gases look like a bird with wings swept back and beak open wide. It is actually a huge star-forming region 7,000 light years away.

## PILLARS OF CREATION

One of the most iconic Hubble Space Telescope images is a close-up of M16, the Eagle Nebula. Hubble zoomed into just the beak of the eagle and astronomers found tall pillars of gas, newly born stars, and the material to make countless solar systems. The picture has been dubbed the *Pillars of Creation*.

## OMEGA MAN

M17 has enough mass to create 30,000 Suns. Also called the Omega Nebula, M17 spans about 40 light years of space and is one of the largest star-forming regions in the galaxy.

## OVERRATED MESSIER OBJECTS

Some miscellaneous Messier objects that are not nebulas, star clusters, or galaxies include M24, M40, and M73. M24 is not a defined structure but merely a dense patch of stars in the Milky Way. M73 is just a small group of stars, the brightest of which make a *V* shape. And M40 is just a double star, which makes you wonder why Messier included it at all.

## MESSIER MYSTERIES

There are several Messier objects that leave astronomers of today scratching their heads. When they search the skies for M47 and M48, they cannot find them in the locations Messier described. M91 is a galaxy in a field full of galaxies. Which one did Messier intend to be M91? And M102 has never been conclusively identified.

## OBSERVATIONAL OBSESSION

In spite of his catalog of deep-space objects, Charles Messier was forever obsessed with comets. According to one account, when his wife was on her deathbed a family friend visited and expressed his deepest sympathy. Messier replied, "Alas! I have discovered a dozen of them; Montaigne had to take away the thirteenth!" He thought the friend's sympathies were because his rival astronomer, Jacques Leibax Montaigne, discovered a comet before Messier, and not because of his ailing wife.

## MESSIER MARATHON

Many amateur astronomers consider it a rite of passage to observe all the Messier objects. Every year in late March, around a new moon, it is technically possible to find all 110 M objects in one night. This is called a Messier marathon. Check with your local astronomy club to see if they are holding a Messier marathon and see if you can join them. Within an hour you'll see star clusters, nebulas, galaxies, and more.

# GALACTIC INTERLUDE

## MILLIONS AND BILLIONS AND TRILLIONS

A galaxy is a massive conglomeration of stars and gas that is bound together by the collective gravity of all the objects. It has a center of mass so strong that it can preside over thousands, millions, billions, or even trillions of stars.

## BARS OF STARS

There are three major types of galaxies that astronomers recognize by shape: spiral, elliptical, and irregular. Spiral galaxies look like flat pinwheels of stars with most stars residing in a central bulge and the spiral arms radiating from it. Some spiral galaxies exhibit a distinct bar of stars near the central bulge and have less distinct arms. This subset of galaxies is referred to as barred spiral galaxies.

## GALAXY GLOSSARY

Elliptical galaxies are huge spheres of stars and look like a ball of light in telescopes. Many of these include the largest known galaxies and have trillions of stars in them.

## DWARF GALAXIES

Galaxies with no definite shape or well-defined edges fall into the category of irregular galaxies. The vast majority of galaxies in the universe are small irregular galaxies called dwarf galaxies. Some dwarf galaxies have several thousand stars in them.

## HIPPY GALAXY

The Andromeda Galaxy is also known as M31 and is composed of approximately 1 trillion stars. M31 is 2.5 million light years away and is the farthest thing you can see with the naked eye under a dark sky. If you look carefully one fall evening, you will see the galaxy as a very tiny cloud near the hip of the constellation Andromeda.

## COLLISION COURSE

The Milky Way and the Andromeda Galaxy are both moving extremely rapidly through the universe, and it appears that they are on a collision course. Andromeda is flying toward the Milky Way at about 250,000 miles per hour. But the two behemoths are so far apart that it will take 3–4 billion years for them to meet.

## A WHOLE NEW MEGAGALAXY

When the Milky Way passes through the Andromeda Galaxy 4 billion years from now the two galaxies will rip whole populations of stars that make up their spiral arms to shreds. Eventually most of the stars will come back together into one megagalaxy. Astronomers are debating the proper name for this future megagalaxy: Milkomeda or Andromeway.

## THINK LOCAL

The Milky Way and its nearest galaxies live in what astronomers call the Local Group. The Local Group includes two barred spiral galaxies (the Milky Way and Andromeda Galaxy), one spiral galaxy (Triangulum Galaxy), and more than fifty-one other irregular and/or dwarf galaxies.

## FALLING INTO THE WHIRLPOOL

A galaxy named M51 is also known as the Whirlpool Galaxy since it looks like it is swallowing up another nearby galaxy. This turns out to be true. M51 is absorbing material from its smaller neighbor, a dwarf galaxy named NGC 5195. The interaction has accentuated the spiral arms and sparked a flurry of star formation.

## SUPER FREAK

The Virgo Cluster is anchored by a freakishly large elliptical galaxy called M87. M87 looks like a globe of light—a sphere of stars jam-packed together. Weighing in at more than 2 trillion suns, it is a freakishly large galaxy lying about 53 million light years from Earth.

## A GALAXY REALLY FAR, FAR AWAY

The farthest galaxy ever seen is called EGSY8p7. Astronomers using the Hubble and Spitzer space telescopes discovered this swarm of stars about 13.2 billion light years from Earth. That means this galaxy formed only about 600 million years after the universe's creation: the Big Bang.

# HOW DO YOU SEE THAT FAR?

## SAY CHEESE!

Cameras are better light detectors than the human eye. Astronomers can take long-exposure photographs to collect much more light and data coming from these distant objects. Astrophotographers can also stack thousands of short-exposure pictures of a Messier object with the use of computer software to draw out details much better than even the keenest eye.

## MEGAPIXEL ENVY

A telescope under construction in Chile, called the Large Synoptic Survey Telescope (LSST), will be getting the most sensitive camera ever made. The 3.2-billion-pixel camera will allow the LSST to record the entire night sky twice a week.

## COLOR ENHANCED

The Hubble Space Telescope takes really pretty, colorful images of deep-space objects. Are they really that colorful? Unfortunately not. Many astronomical images are Photoshopped in order to bring out contrast, faint structures, and individual elements and gases. So if you were flying close to the Orion Nebula, it would not shine with all the colors of the rainbow.

# BEYOND THE MESSIER CATALOG

### HERSCHEL'S HOT LIST

English astronomer William Herschel observed so many deep-sky objects that he made his own catalog in 1786, called the *Catalogue of Nebulae and Star Clusters*. Herschel's catalog greatly expanded on the one that Charles Messier had created earlier and included about 1,000 deep-space gems. Herschel added another 1,500 can't-miss objects to his catalog over the course of his life.

### NEITHER SEAHORSE NOR SPACEHORSE

You've probably seen a picture of the Horsehead Nebula (also designated in another catalog as Barnard 33), which is part of the constellation Orion. It looks like a silhouette of a black stallion in front of a sea of pinkish gas, but in reality it's another star-forming region. The darker head of the horse is filled with so much dust that it blocks the view of the ionized hydrogen gas coming from behind it. The nearby massive star Sigma Orionis is the one supplying the gas that outlines the horsehead.

### THE SHAPE OF NOTHINGNESS

NGC 1999 looks like a black keyhole surrounded by iridescent gases that reflect the light of one star. Astronomers first assumed that the inky center of this nebula was caused by dark material blocking the brighter gases behind it (like the Horsehead Nebula). Now it turns out that the keyhole is truly empty space. Why is there nothing there? And how did the gases form such an exquisite shape? Astronomers don't know.

## CALDWELL AND THE DOUBLE CLUSTER

Sir Patrick Alfred Caldwell-Moore created the Caldwell Catalog, a collection of 109 objects that make excellent targets for amateur astronomers and their telescopes. One such target is the Double Cluster, located between the constellations Perseus and Cassiopeia. The Double Cluster looks exactly as you would expect: two collections of hundreds of stars, side by side. These open clusters lie about 7,500 light years from Earth but can be seen with the naked eye under ideal and dark sky conditions.

## WHAT MESSIER MISSED

The Caldwell Catalog included many deep-sky objects that Charles Messier could not observe from his observatory in France. The star clusters Omega Centauri, 47 Tucanae, and the Jewel Box; bright galaxies like Centaurus A; and nebulas like Carina and the Tarantula can best be seen from the Southern Hemisphere and made Moore's list.

## THE VIEW FROM DOWN SOUTH

Easily visible to the naked eye in the Southern Hemisphere, the globular cluster Omega Centauri covers almost the same amount of sky as the full moon. This is the largest globular cluster in the Milky Way. It is 15,000 light years away, 170 light years wide, has over 100 million stars, and weighs in at more than 4 million Suns. Astronomers continue to debate the origin of this behemoth star cluster and wonder if it was perhaps the remnants of a galaxy that got too close to our Milky Way.

## LACAILLE'S TOUCAN

The globular cluster called 47 Tucanae can be seen with the naked eye, but the French astronomer Nicolas-Louis de Lacaille is said to have "discovered" it during an expedition to the Cape of Good Hope, South Africa. Lacaille noted 47 Tucanae in 1751 when he mistook it for a comet. It is, in fact, the second-brightest globular cluster in the Milky Way (behind Omega Centauri), is about 16,700 light years from Earth, and has a mass of about 700,000 Suns.

## THE CARINA COMBINATION

The Carina Nebula is a humongous region filled with stars under construction, newly formed star clusters, and one stellar behemoth nearing the end of its life. This strange combination of unborn, newly born, and dying stars 7,500 light years away is visible from the Tropics and further south.

## THERE SHE BLOWS

The most famous star inside the Carina Nebula is called Eta Carinae, a star about 4 million times brighter than the Sun. If it were closer than 7,500 light years away from Earth, Eta Carinae would be one of the brightest stars in the sky. And it actually was for four days in 1843. That year, Eta Carinae erupted and brightened to the point that it was equal to Alpha Centauri, the third-brightest star in the sky.

## SAY BUB-BYE

The Hubble Space Telescope has imaged two great bubbles of gas that are shooting out of Eta Carinae. The giant star is in its final death throes and may make a supernova explosion any day. Astronomers are anxiously awaiting this dramatic demise of a super star.

## LONE WOLF

WR 25, contained within the Carina Nebula, is a rare type of massive object (called a Wolf-Rayet star) that is the most luminous known star in the galaxy. It shines 6.3 million times brighter than the Sun. WR 25 cannot be seen with the naked eye, however. It is surrounded by thick envelopes of gas that make it appear dimmer than it is.

## INSIDE THE JEWEL BOX

The Jewel Box, in the constellation Crux, AKA the Southern Cross, is an open cluster similar to the Pleiades. At about 14 million years old, the 100 stars within the Jewel Box are some of the most recently formed stars in the galaxy.

## SEVEN SISTERS OF THE SOUTHERN SKIES

Theta Carinae, an open cluster, is nicknamed the Southern Pleiades. At about 479 light years from Earth, the sixty stars that make up this cluster are similar in distance to their northern cousins. But they are a much fainter gathering of stars and look best when viewed through a pair of binoculars.

## ALMOST ABSOLUTE ZERO

The Boomerang Nebula looks like a star entering the next phase of its life and is just beginning to turn into a planetary nebula. Astronomers have measured the coldest temperature in the universe inside this nebula: 1° Kelvin, or one degree above absolute zero.

# BLACK HOLES

## MYSTERY OF DEEP SPACE

A black hole is a place in space where gravity is so intense that nothing can escape it, not even light. So what happens to the matter inside a black hole? Since you can't see this stuff inside, no one knows for certain.

## BLACK HOLE-FREE

Earth will never naturally turn into a black hole. You'd have to condense everything on the entire planet into the size of a marble. Take an 8,000-mile-wide object and crush into a ½-inch ball. That's the kind of density within a black hole. It can happen with stars but won't happen to Earth.

## SO FAR YOU'RE FINE

Earth will not be swallowed up by a black hole. There are no black holes near it. The closest ones are thousands of light years away. And the supermassive black hole in the center of the Milky Way is about 27,000 light years from Earth. Plus a black hole is really small, and so is its gravitational reach. Many black holes are smaller than Earth, and some are just six miles across. You have to be really close to a black hole to fall in, so don't worry.

## SUPERBIG SAGITTARIUS A*

The supermassive black hole in the center of the galaxy is called Sagittarius A*. It is estimated to contain 4.3 million times the mass of the Sun.

## HARD TO PICTURE

Astronomers don't have any pictures of black holes. Think about it: What color is space? Black. What color are black holes? Black. Yes, that makes them tough to photograph and study in visible light.

## IF THE NAME FITS

Black holes exist in some form because astronomers can see what they do to things around them. They move objects like stars and they excite x-rays. Once you know where these black holes are located, you can study them in all wavelengths of light and "see" beyond what the eye can see.

## CYGNUS X-1

The first black hole was discovered in 1964 during a rocket flight. The rocket detected an unusual amount of x-rays coming from a seemingly empty region of the constellation Cygnus. It turned out to be a black hole about 6,000 light years away, and it is called Cygnus X-1. The closest black hole to Earth is likely A0620-00, another x-ray source that lies about 3,000 light years from Earth.

## WHEN NERDS BET

Astronomers Stephen Hawking and Kip Thorne made a bet about Cygnus X-1, the first confirmed black hole. Was it really a black hole? Hawking said no. Thorne said yes. As more evidence pointed toward yes, Hawking conceded the bet and, as he wrote in his book *A Brief History of Time*, "paid the specified penalty, which was a one-year subscription to *Penthouse*, to the outrage of Kip's liberated wife." If Hawking won, he would have received a four-year subscription to *Private Eye* magazine.

# CHAPTER 10

# DEEP THOUGHTS ABOUT SPACE

*Astronomy's Wow Factor*

Once upon a time, about 13.8 billion years ago, everything was in one place. The universe was a singularity, an infinitely small spot of infinite density and temperature. Then, bang! The singularity exploded and the universe came into being. Scientists have mounds of evidence to back up this Big Bang theory, but many deep questions remain.

Why is there something and not nothing? Why did the singularity go "bang" and create the universe? Could it have easily just remained as nothing? Astronomers have a good handle on what occurred just after the Big Bang, but physics breaks down at the moment of creation. What happened to trigger everything? Can that question even be answered?

When presented with a particular scenario, astronomers invariably ask, "What came before that?" What happened before the Big Bang? What created the singularity? Was it there all the time or was there a time when there was nothing? And if so, how do you get something from nothing? These are the deep thoughts that keep astronomers up at night and make everyone say, "Wow!"

For your final journey you'll trek through space, time, and imagination and explore the details beyond the facts and try to answer the biggest questions of them all.

# CONSIDER THE UNIVERSE

## CRUNCH TIME?

The fate of the universe has been debated by astronomers for the past century. Originally many astronomers believed that the universe's expansion would slow over time. Gravity would act on the matter and energy created 13.8 billion years ago and would draw them back together in a reverse Big Bang—a Big Crunch. The universe would once again become a singularity. From this moment a new Big Bang could occur and the process could start all over again. If this were true, how many times has the universe been created? Which universe are you living in, the first or merely the latest in a series?

## THE BIG FREEZE

Current evidence shows that the universe is expanding at such a rate that gravity will not be able to pull everything back together again. That means the universe will continue to expand, galaxies will spread out, stars will become more distant from each other, and things will get colder. That means there will be no Big Crunch. Eventually, if this scenario plays out, the temperature of the universe will reach absolute zero and won't have any light, heat, or energy. This is called the Big Freeze theory.

## THE END OF THE UNIVERSE

There are many factors that may affect the timeline of the Big Freeze. Stars could stop forming in about 100 trillion years. And the moment when the universe is cold, dark, and empty may still be one googol years away (that's a one followed by 100 zeros, or 10,000,000,000,000,000,00 0,000,000,000,000,000,000,000,000,000,000,000,000,000,00 0,000,000,000,000,000,000,000,000,000,000 years).

## GASES HAVE MASSES

Everything that has mass has gravity. Earth, the Sun, your cat, even the air you breathe all have mass. Although stars like the Sun and planets like Jupiter are made up almost entirely of gases, they still have a tremendous amount of mass. That's right, gases have masses.

## A FORCEFUL READ

You are pulled by the gravity of the Moon, Sun, and planets. However, since the force of gravity greatly diminishes with distance, these objects have an extremely weak influence on you. In fact, this book that you are holding, by its sheer proximity, exerts a greater gravitational force on you than any of the planets in the solar system.

## WHEN WORLDS ALIGN

When planets align, nothing happens on Earth. It's possible to have two, three, or four planets appear to line up in the nighttime sky. This is because all of the planets are in just about the same plane in the solar system. To a varying degree, they always line up. When all five naked-eye planets are visible at one time, there is no danger to Earth. It's just cool to see!

## GET IN LINE

The eight planets will never line up perfectly. Their orbits are too different from one another: different periods, different inclinations, and different eccentricities. In March 2854, the eight planets will be within 45 degrees of each other, but that means they're still spread across ¼ of the sky as viewed from Earth.

## ZERO PERCENT SUCCESS RATE

Every astronomical doomsday prediction has been wrong. There wasn't a mass extinction when the planets lined up in 2003. Earth did not end on December 21, 2012, when the Mayan calendar supposedly "ran out." Halley's comet didn't kill anyone with cyanide in 1910, and Comet Elenin didn't crash into Earth in 2011. Blood moons and eclipses have no effect on Earth. These and many more astronomical events led people to broadcast unfounded predictions for global Armageddon. Zero percent of them transpired.

## COSMOPHOBIA

Cosmophobia is an irrational fear of the universe. American astronomer David Morrison coined this term to describe the condition that many people exhibited through correspondence on the *Ask an Astrobiologist* blog on NASA's website. These folks were genuinely afraid that something will fall from the sky, that Earth will someday stop spinning or flip over, or that the Sun will either burn out or burn too brightly and cook us.

## PULLED IN EVERY DIRECTION

If you were in the very center of Earth (and you could somehow survive the 10,000°F heat), you would be weightless. The enveloping Earth would pull on you equally in all directions. You'd float.

## YOU'LL KNOW IT WHEN YOU SEE IT

Stars are light years away. So the light you see twinkling at night left those stars years, centuries, or millennia ago. What do those stars really look like? They've most certainly aged, moved, gotten bigger, and some of them might have died in tremendous supernova explosions. Astronomers cannot predict if a star has already exploded. You won't know when a star goes kablooey until the kablooey reaches Earth at the speed of light (186,000 miles per second).

## THE SPEED OF LIGHT

Light travels at 186,000 miles per second. A camera flash from your cell phone would take about 1.3 seconds to reach the Moon and about 500 seconds to reach the Sun (not that it needs any extra light!).

## FASTER THAN THE SPEED OF LIGHT

Nothing is faster than the speed of light, right? Humorist and author Kurt Vonnegut suggested an exception to this fact in his 1997 novel **Timequake**. For example, look at a star. Then look at another star. Your eyes just perceived objects separated by many light years. Your brain processed that widely separated information in less than a second. There is something faster than light: human consciousness.

## THE WAY IT WAS

Galaxies are millions of light years away. The Andromeda Galaxy, for example, is 2.5 million light years away. That means the light you see in a telescope left that galaxy 2.5 million years ago. Likewise, astronomers in the Andromeda Galaxy see Earth as it was 2.5 million years ago.

## T-REX REALITY TV

What could aliens on distant planets see of Earthlings? It depends on how far away they were. If a planet was 50 light years away, they could watch Super Bowl I. If they were 100 light years away, they could see World War I. If they were 65 million light years away, they could watch the last days of the dinosaurs.

## TIMELESS CONSTELLATIONS

The stars in the Milky Way rapidly move through space. However, they are so far away from Earth that your perspective on them barely changes. That means almost every constellation you see today looked about the same as it did when, in 1492, Columbus sailed the ocean blue, or when ancient Babylonians stargazed thousands of years ago.

## THE BIG SPATULA

If you looked at the stars 50,000 years from now, you'd see a change. In fact, you'd hardly recognize many constellations. The stars in the Big Dipper, for instance, will shift among themselves and look like a "Big Spatula" by the year A.D. 75000.

## MIGRATING PLANETS

Mercury, Venus, Earth, Mars, Jupiter, Saturn, Uranus, and Neptune. This has not always been the order of the planets. Astronomers now believe that in the distant past planets migrated to their current positions. From where? Evidence strongly suggests that Uranus and Neptune switched places as the seventh and eighth planets. Neptune was, at one time, closer to the Sun than Uranus.

## A PLANET ON PAPER

American astronomers Mike Brown and Konstantin Batygin have discovered another planet in the solar system: Planet Nine. One slight problem—no one has seen it yet. Mathematically, Planet Nine should be there. The presence of a planet almost as big as Neptune in the far reaches of the solar system may explain the odd movements of Kuiper belt objects. Some astronomers are searching space to become the first to spy this deep, dark world—if it exists.

## A HOME FOR MR. SPOCK

In the 1800s, several astronomers believed that there was another planet closer to the Sun than Mercury. A few astronomers even claimed to have observed it going in front of the Sun. They named it (wait for it) . . . Vulcan. Many astronomers looked for planet Vulcan for decades, but other than one very questionable sighting in 1878, no one ever saw it again. The mythical planet was most likely the result of astronomers staring too long at the Sun and hoping and wishing to discover a new planet. Vulcan does not exist.

## WHAT THE GREEKS KNEW

In the third century B.C., a Greek astronomer named Aristarchus of Samos attempted the seemingly impossible—to measure the size of and distance to the Moon and Sun. Through some clever geometric tricks he came up with an extremely accurate distance to the Moon. The Sun was another story. Aristarchus determined that the Sun was 5 million miles away. Although the real figure is closer to 93 million miles, his findings led Aristarchus to a conclusion so radical that no one else in the ancient world shared it: Earth goes around the Sun.

## WHAT'S WOBBLING?

Around 130 B.C., a Greek astronomer named
Hipparchus wrote about an astronomical cycle
called precession. After studying centuries'
worth of Greek star charts and data from
past Babylonian astronomers, Hipparchus
could see for himself that the positions of
the stars had shifted. He figured out that
the entire sky made one wobble every
26,000 years. Although, in fact, it is
Earth doing the wobbling, Hipparchus
was correct about the length of the
wobble: 26,000 years.

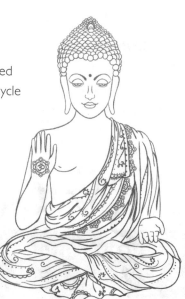

## TIME AND PLACE

The ancient Inca of Peru have a word, *pacha*, which has many intriguing
translations. It could mean "the world at this moment," or possibly taken
in a larger sense "the cosmos as it exists." Pacha links two key astronomical concepts in the same way that Albert Einstein did in the early
twentieth century: space and time. Einstein proposed that the universe
you experience is relative to space and time and thus should be thought
of as one thing: space-time.

## THINK BIG

Ancient Hindu mythology describes the life of the universe spanning over 100 trillion years. In a late second millennium B.C. text called the *Rigveda*, philosophers asked the deep questions about the origins of the universe: "What was concealed? And where? And in whose protection? Who really knows? Who can declare it? When was it born, and when came this creation?" It goes on to describe the universe they saw as only the latest universe, the latest cycle of life, death, and rebirth lasting eons upon eons.

## SEVENTEENTH-CENTURY BLASPHEMY

In 1600, an Italian friar named Giordano Bruno was burned at the stake with a spike through his tongue for avowing (among other heretical things) that Earth went around the Sun. Bruno also said that the stars he saw at night were other suns, that these suns may have planets orbiting them, and on those planets other life forms may exist.

## THE LIFE AND DEATH OF AN ECCENTRIC BILLIONAIRE ASTRONOMER

Sixteenth-century Danish nobleman and astronomer Tycho Brahe was the greatest naked-eye astronomer and an interesting character. His colorful past included being kidnapped by his uncle, being given an island, constructing a state-of-the-art and gorgeous observatory, being evicted from his island only to have his observatory destroyed by the islanders after his departure, and wearing a metal prosthetic nose after having the tip of his own nose cut off during a mathematics-inspired duel. Brahe died in 1601 as a result of a burst bladder after he held his pee too long.

## ASTROLOGY PAYS THE BILLS

Renaissance astronomers like Brahe, Galileo, Kepler, and Copernicus also practiced astrology. The practice was a holdover from more superstitious times and was quite lucrative. However, one generation later, astrology all but died out. It was considered by scientists of the Enlightenment to be complete rubbish.

## CON ARTISTS AND CRIMINALS

The astrology of today re-emerged from the Enlightenment through a series of con artists and criminals in early twentieth-century England and America. Astrology was akin to fortunetelling and treated as a punishable crime in England. Foreign practitioners of astrology were even deported for swindling people out of their money.

# ALIENS, DARK MATTER, AND THE MULTIVERSE

### THE FERMI PARADOX

There is no concrete proof that aliens exist, and in the 1950s, Italian physicist Enrico Fermi posed some skeptical questions. If there are aliens, then where are they already? Why haven't they visited or communicated with us? If the universe is so old and alien civilizations could advance to technological heights far greater than ours, what's keeping them? This is the Fermi paradox.

### A SERIES OF FORTUNATE EVENTS

Astronomers ask a lot of questions about alien life. Maybe intelligent life is incredibly rare. Maybe we are alone in the universe. Are we at the end of a fantastically rare series of events that made a universe, created a galaxy, formed the Sun and planets, and made one planet—and one planet alone—perfect for life to form and survive? If even one variable in our creation story was altered, would we still be here today?

### AN ALIEN FORMULA

American astronomer Frank Drake proposed an equation to statistically measure the chances of extraterrestrial life existing in the universe. The Drake equation asks: What is the rate of new stars being born? What percentage of those stars has planets? How many of these planets might support life? What percentage of those planets will actually have life forms? What percentage of those planets will have intelligent life forms inhabiting them? What percentage of those will be able to communicate or cross space to Earth? And how long will such a civilization last? Unless the numbers for any of these factors is zero, there is a statistical probability that somewhere beyond Earth intelligent life exists.

## LISTENING CLOSELY

The SETI (search for extraterrestrial intelligence) Institute is a collaborative effort of scientists around the world whose purpose is to detect the presence of alien life. The most prominent investigations use large radio telescopes to listen for signals that originated from alien civilizations. Someday, if they tune in long enough, they may hear a clear message from another world far, far away.

## A WILD ALIEN CHASE

Today, some scientists are critical of SETI. They believe that the current technology available to scientists is not able to detect alien transmissions. If aliens sent a signal, would scientists even be able to recognize it? Astronomers debate whether or not signals can actually be sent across hundreds of light years. The signals may decay before reaching another inhabited planet, which might explain why they haven't heard any.

## ROCKING OUT TO THE GOLDEN RECORD

American astronomer Carl Sagan supervised the design and construction of messages sent to aliens. They were attached to the Voyager 1 and Voyager 2 spacecraft in the form of a long-play golden record (one on each spacecraft). Any aliens who find the spacecraft could play the record (record player included) and hear sounds from Earth like rainstorms, frogs, and thunder; see 115 images; and hear twenty-seven songs from around the world.

## WHAT ARE THEY MISSING?

Many astronomers believe that the majority of the universe is currently invisible to their telescopes. The matter of the planets, stars, and galaxies—what is visible—makes up only 4 percent of the total mass of the universe. Where is the other 96 percent? It is possibly hidden from current technology in the forms of dark matter and dark energy.

## COME TO THE DARK SIDE

Although dark matter and dark energy have not been directly observed, the majority of astronomers think that their presence explains peculiar movements noted in ordinary matter. Galaxies spin faster at their edges than they should. Galaxies are moving away from Earth at increasing velocities when they shouldn't. What is pulling or pushing them? Dark matter and dark energy are two possibilities.

## CONSIDER THE POSSIBILITIES

Do dark energy and dark matter exist? Detractors to this theory of invisible matter and energy argue that there is not enough data about the visible universe to jump to this conclusion. Maybe the measurements of the farthest objects in the universe are not as accurate as scientists think. Perhaps gravity does not work exactly as the scientists postulate, and a slight modification to the laws of gravity could account for any peculiar movements on the largest scales.

## DRIFTING AWAY

Almost all of the galaxies in the universe are moving away from the Milky Way. As the universe expands it takes these galaxies on a ride away from the Milky Way at an accelerating pace. If galaxies keep accelerating away and reach the speed of light, they will not be visible anymore.

## WHAT'S THE LIMIT?

The universe is everything that you can see. That includes all of the planets, all of the stars, all of the galaxies, and all the light, energy, matter, gas, and mass. In short, the universe is everything. What lies outside the universe? Since you cannot see it, no one knows.

## HOW IT APPEARS

Is the universe infinite? You can look at it two ways. Yes, the universe incorporates everything. It has no boundaries, no ends. Therefore, it is infinite. However, from an astronomer's perspective inside the universe, it appears to have measurable limits. The farthest things in the universe look like they're 13.8 billion light years away. So at first glance the universe appears to be about 27.6 billion light years across.

## BUT WHAT IS IT REALLY LIKE?

Since the Big Bang, the universe has expanded. Space itself grows. When astronomers factor in the age of the objects and add it to the expected expansion rate, the universe actually measures about 92 billion light years across.

## SCIENCE FICTION VERSUS SCIENCE FACT

Some scientists claim that there are multiple universes in existence. Perhaps this universe is just one of an infinite number of universes. With infinite universes there may be an infinite number of Earths with infinite people experiencing life in various ways and forms. But since there is absolutely no evidence to support such a thing, the multiverse idea, at this time, is complete science fiction.

## THERE ARE NO OTHER UNIVERSES

Will astronomers ever see another universe? This prospect is difficult to achieve, and the idea conflicts with the current definition of the universe. If the universe is everything you can see, there are no other universes. If astronomers someday detect another universe, the moment they see it, it ceases to be another universe. It immediately becomes part of everything. It becomes part of *the* universe.

# ABOUT THE AUTHOR

Dean Regas is the astronomer for the Cincinnati Observatory and the cohost of the syndicated television program *Star Gazers*, which airs nightly on PBS stations around the country. He is a contributing editor to *Sky & Telescope* magazine and frequent contributor to *Astronomy* magazine, and you can often hear Dean on the radio program *Science Friday* with Ira Flatow. He lives in a colorful house on one of Cincinnati, Ohio's seven hills.